I0489264

Evapotranspiration from Wetland and Open-Water Sites at Upper Klamath Lake, Oregon, 2008–2010

By David I. Stannard, Marshall W. Gannett, and Danial J. Polette, U.S. Geological Survey; Jason M. Cameron, Bureau of Reclamation; M. Scott Waibel, U.S. Geological Survey; and J. Mark Spears, Bureau of Reclamation

Prepared in cooperation with the Bureau of Reclamation

Scientific Investigations Report 2013–5014

U.S. Department of the Interior
U.S. Geological Survey

U.S. Department of the Interior
KEN SALAZAR, Secretary

U.S. Geological Survey
Suzette M. Kimball, Acting Director

U.S. Geological Survey, Reston, Virginia: 2013

For more information on the USGS—the Federal source for science about the Earth, its natural and living resources, natural hazards, and the environment, visit http://www.usgs.gov or call 1–888–ASK–USGS.

For an overview of USGS information products, including maps, imagery, and publications, visit http://www.usgs.gov/pubprod

To order this and other USGS information products, visit http://store.usgs.gov

Contents

Figures

Tables

Conversion Factors

Inch/Pound to SI

Multiply	By	To obtain
Length		
inch (in.)	2.54	centimeter (cm)
inch (in.)	25.4	millimeter (mm)
foot (ft)	0.3048	meter (m)
mile (mi)	1.609	kilometer (km)
yard (yd)	0.9144	meter (m)
Area		
acre	4,047	square meter (m^2)
acre	0.4047	hectare (ha)
acre	0.4047	square hectometer (hm^2)
acre	0.004047	square kilometer (km^2)
square foot (ft^2)	0.09290	square meter (m^2)
section (640 acres or 1 square mile)	259.0	square hectometer (hm^2)
square mile (mi^2)	259.0	hectare (ha)
square mile (mi^2)	2.590	square kilometer (km^2)
Volume		
gallon (gal)	0.003785	cubic meter (m^3)
million gallons (Mgal)	3,785	cubic meter (m^3)
cubic inch (in^3)	16.39	cubic centimeter (cm^3)
cubic foot (ft^3)	0.02832	cubic meter (m^3)
cubic yard (yd^3)	0.7646	cubic meter (m^3)
acre-foot (acre-ft)	1,233	cubic meter (m^3)
Flow rate		
acre-foot per day (acre-ft/d)	0.01427	cubic meter per second (m^3/s)
acre-foot per year (acre-ft/yr)	1,233	cubic meter per year (m^3/yr)
foot per second (ft/s)	0.3048	meter per second (m/s)
cubic foot per second (ft^3/s)	0.02832	cubic meter per second (m^3/s)
million gallons per day (Mgal/d)	0.04381	cubic meter per second (m^3/s)
inch per day (in/d)	25.4	millimeter per day (mm/d)
inch per year (in/yr)	25.4	millimeter per year (mm/yr)
mile per hour (mi/h)	0.4470	meter per second (m/s)
Mass		
ounce, avoirdupois (oz)	28.35	gram (g)
pound, avoirdupois (lb)	0.4536	kilogram (kg)
Pressure		
atmosphere, standard (atm)	101.3	kilopascal (kPa)
bar	100	kilopascal (kPa)
inch of mercury at 60 °F (in. Hg)	3.377	kilopascal (kPa)
pound-force per square inch (lbf/in^2)	6.895	kilopascal (kPa)
pound per square foot (lb/ft^2)	0.04788	kilopascal (kPa)
pound per square inch (lb/in^2)	6.895	kilopascal (kPa)
Density		
pound per cubic foot (lb/ft^3)	16.02	kilogram per cubic meter (kg/m^3)
pound per cubic foot (lb/ft^3)	0.01602	gram per cubic centimeter (g/cm^3)
Energy		
kilowatthour (kWh)	3,600,000	joule (J)
calorie (cal)	4.187	joule (J)

Temperature in degrees Celsius (°C) may be converted to degrees Fahrenheit (°F) as follows:

°F=(1.8×°C)+32

Temperature in degrees Fahrenheit (°F) may be converted to degrees Celsius (°C) as follows:

°C=(°F-32)/1.8

Vertical coordinate information is referenced to the National Geodetic Vertical Datum of 1929 (NGVD 29).

Horizontal coordinate information is referenced to the North American Datum of 1927 (NAD 27).

Altitude, as used in this report, refers to distance above the vertical datum.

SI to Inch/Pound

Multiply	By	To obtain
Length		
centimeter (cm)	0.3937	inch (in.)
millimeter (mm)	0.03937	inch (in.)
meter (m)	3.281	foot (ft)
kilometer (km)	0.6214	mile (mi)
meter (m)	1.094	yard (yd)
Area		
square meter (m^2)	0.0002471	acre
hectare (ha)	2.471	acre
square hectometer (hm^2)	2.471	acre
square kilometer (km^2)	247.1	acre
square meter (m^2)	10.76	square foot (ft^2)
square hectometer (hm^2)	0.003861	section (640 acres or 1 square mile)
hectare (ha)	0.003861	square mile (mi^2)
square kilometer (km^2)	0.3861	square mile (mi^2)
Volume		
cubic meter (m^3)	264.2	gallon (gal)
cubic meter (m^3)	0.0002642	million gallons (Mgal)
cubic centimeter (cm^3)	0.06102	cubic inch (in^3)
cubic meter (m^3)	35.31	cubic foot (ft^3)
cubic meter (m^3)	1.308	cubic yard (yd^3)
cubic meter (m^3)	0.0008107	acre-foot (acre-ft)
Flow rate		
cubic meter per second (m^3/s)	70.07	acre-foot per day (acre-ft/d)
cubic meter per year (m^3/yr)	0.000811	acre-foot per year (acre-ft/yr)
meter per second (m/s)	3.281	foot per second (ft/s)
cubic meter per second (m^3/s)	35.31	cubic foot per second (ft^3/s)
cubic meter per second (m^3/s)	22.83	million gallons per day (Mgal/d)
millimeter per day (mm/d)	0.03937	inch per day (in/d)
millimeter per year (mm/yr)	0.03937	inch per year (in/yr)
meter per second (m/s)	2.237	mile per hour (mi/h)
Mass		
gram (g)	0.03527	ounce, avoirdupois (oz)
kilogram (kg)	2.205	pound avoirdupois (lb)
Pressure		
kilopascal (kPa)	0.009869	atmosphere, standard (atm)
kilopascal (kPa)	0.01	bar
kilopascal (kPa)	0.2961	inch of mercury at 60°F (in. Hg)
kilopascal (kPa)	0.1450	pound-force per square inch (lbf/in^2)
kilopascal (kPa)	20.88	pound per square foot (lb/ft^2)
kilopascal (kPa)	0.1450	pound per square inch (lb/in^2)
Density		
kilogram per cubic meter (kg/m^3)	0.06242	pound per cubic foot (lb/ft^3)
gram per cubic centimeter (g/cm^3)	62.4220	pound per cubic foot (lb/ft^3)
Energy		
joule (J)	0.000000278	kilowatthour (kWh)
joule (J)	0.2388	calorie (cal)
Power		
watt (W)	0.001341	horsepower (hp)

Temperature in degrees Celsius (°C) may be converted to degrees Fahrenheit (°F) as follows:

°F=(1.8×°C)+32

Temperature in degrees Fahrenheit (°F) may be converted to degrees Celsius (°C) as follows:

°C=(°F-32)/1.8

Vertical coordinate information is referenced to the National Geodetic Vertical Datum of 1929 (NGVD 29).

Horizontal coordinate information is referenced to the North American Datum of 1927 (NAD 27).

Altitude, as used in this report, refers to distance above the vertical datum.

Acknowledgments

This investigation was funded largely by the Science and Technology Program of the Bureau of Reclamation Research and Development Office and from the National Research Program of the U.S. Geological Survey. Most of the specialized eddy-covariance and meteorological equipment was purchased by the Bureau of Reclamation Klamath Basin area office. Thanks are extended to the U.S. Fish and Wildlife Service for the reconnaissance efforts of Dave Mauser and Tim Mayer and for cooperation in providing access to the Upper Klamath National Wildlife Refuge. Thanks are also extended to Amanda Garcia and David Sumner of the U.S. Geological Survey for their helpful reviews of the manuscript. Finally, thanks go to Michael Johnson of the U.S. Geological Survey for his helpful discussion on specialized supports for deployment of the wetland eddy-covariance stations.

Evapotranspiration from Wetland and Open-Water Sites at Upper Klamath Lake, Oregon, 2008–2010

By David I. Stannard[1], Marshall W. Gannett[1], Danial J. Polette[1], Jason M. Cameron[2], M. Scott Waibel[1], and J. Mark Spears[2]

Abstract

Water allocation in the Upper Klamath Basin has become difficult in recent years due to the increase in occurrence of drought coupled with continued high water demand. Upper Klamath Lake is a central component of water distribution, supplying water downstream to the Klamath River, supplying water for irrigation diversions, and providing habitat for various species within the lake and surrounding wetlands. Evapotranspiration (ET) is a major component of the hydrologic budget of the lake and wetlands, and yet estimates of ET have been elusive—quantified only as part of a lumped term including other substantial water-budget components. To improve understanding of ET losses from the lake and wetlands, measurements of ET were made from May 2008 through September 2010. The eddy-covariance method was used to monitor ET at two wetland sites continuously during this study period and the Bowen-ratio energy-balance method was used to monitor open-water lake evaporation at two sites during the warmer months of the 3 study years. Vegetation at one wetland site (the bulrush site) consists of a virtual monoculture of hardstem bulrush (formerly *Scirpus acutus*, now *Schoenoplectus acutus*), and at the other site (the mixed site) consists of a mix of about 70 percent bulrush, 15 percent cattail (*Typha latifolia*), and 15 percent wocus (*Nuphar polysepalum*). Measured ET at these two sites was very similar (means were ±2.5 percent) and mean wetland ET is computed as a 70 to 30 percent weighted average of the bulrush and mixed sites, respectively, based on community-type distribution estimated from satellite imagery.

Biweekly means of wetland ET typically vary from maximum values of around 6 to 7 millimeters per day during midsummer, to minimum values of less than 1 mm/d during midwinter. This strong annual signal primarily reflects life-cycle changes in the wetland vegetation, and the annual variation of radiative input to the surface and resulting temperature. The perennial vegetation begins each growing season submerged, emerges from the dead litter mat around late May or early June, reaches a maximum height of about 2.2 meters (m) during summer, senesces in October, and subsequently lodges over, contributing to the dead litter mat from previous years. Hydroperiods last about 5 to 6 months, typically beginning in January or February and ending in July or August, and have a minor influence on the annual ET cycle. These hydroperiods result from lake levels that typically vary about 1.3 m, from around 0.6 to 0.9 m above the wetland surface, to around 0.4 to 0.7 m below the wetland surface.

An estimate of 3-year annual wetland ET, made by substituting early- and late-season data measured during 2009 for the missing periods in early 2008 and late 2010, is 0.938 meter per year (m/yr). Daily values of alfalfa-based reference ET (ET_r) were retrieved from the Bureau of Reclamation AgriMet Web site (http://www.usbr.gov/pn/agrimet/index.html) and are aggregated into biweekly, annual, and 3-year values (for consistency, the 3-year values are also computed using substitute data from 2009 for early 2008 and late 2010). These ET_r values are computed from weather data measured at the nearby Agency Lake weather station (AGKO), and are based on the assumption that the alfalfa crop is green and vigorous year-round. The 3-year value of ET_r is 1.145 m/yr, about 22 percent greater than wetland ET. A comparison of 2008–2010 alfalfa and pasture growing season actual ET with wetland ET is made using data from the more distant Klamath Falls AgriMet weather station (KFLO) because actual alfalfa and pasture ET are not computed for the AGKO site. During the 190-day average alfalfa growing season, wetland ET (0.779 m) is about 7 percent less than alfalfa ET (0.838 m). During the 195-day average pasture growing season, wetland ET (0.789 m) is about 18 percent greater than pasture ET (0.671 m). Assuming alfalfa and pasture ET are equal to wetland ET during the non-growing season, annual estimates become 0.997 m, 0.938 m, and 0.820 m from alfalfa, wetland, and pasture, respectively.

[1] U.S. Geological Survey.

[2] Bureau of Reclamation.

Wetland crop coefficients ($K_c = ET/ET_r$) are computed at daily, biweekly, and annual time steps. Approximate formulas are given to estimate daily values of growing season K_c, thereby allowing computation of daily growing season ET using ET_r from the AGKO weather station. Biweekly values of growing season K_c are computed from ensemble average values of ET and ET_r during the 3 study period growing seasons, and a single, mean K_c is computed for the non-growing season. Together, these provide relatively accurate estimates of biweekly ET during the study ($RMSE = 0.396$ and 0.347 mm/d, $r^2 = 0.962$ and 0.971 at the bulrush and mixed sites, respectively). A fourth-order polynomial fit of the biweekly growing season values to day of year provides a more automated form of ET computation.

Measured ET at the bulrush wetland site during the current study compares very closely with growing-season ET estimated during a study in 1997 at nearly the same location. During the earlier study, ET was measured four times, using eddy covariance for 1- to 2-day periods, and was estimated between measurement periods using a Penman-Monteith model, calibrated to the measurements. Differences between time series of ET from the two studies are similar to interannual differences within the current study. Compared to the 1997 study, the current study measured larger ET rates in early summer and smaller rates in late summer, resulting in very similar growing-season totals.

A study conducted in 2000 estimated ET from nearby fallowed cropland, using the Bowen-ratio energy balance method supplemented with Priestley-Taylor and crop-coefficient ET modeling. Seasonal timing of ET from three different crop types varied considerably, but growing-season totals were remarkably similar, at 0.435 ± 0.009 m. Wetland ET measured during the current study, evaluated over the same growing season was 0.718 m, or about 65 percent greater than the fallowed cropland ET.

Open-water evaporation from Upper Klamath Lake was measured at two locations during the warmer months of 2008–2010 using the Bowen-ratio energy balance method. Measured rates were in general agreement with those measured in 2003 using the same method. Open-water evaporation and wetland ET were nearly equal during late June through early August, when wetland vegetation was green and abundant. As expected, open-water evaporation consistently exceeded wetland ET during late summer, as wetland ET responded to vegetation senescence while open water evaporation responded to extra available energy in the form of heat previously stored in the lake. Overall, open-water evaporation was 20 percent greater than wetland ET during the same period.

Introduction

Upper Klamath Lake is a primary source of irrigation water for the Bureau of Reclamation's Klamath Reclamation Project, which delivers irrigation water to approximately 210,000 acres of croplands in south-central Oregon and northern California. The lake also provides habitat for two species of fish listed as endangered under the Endangered Species Act, the short-nose sucker (*Chasmistes brevirostris*) and the Lost river sucker (*Deltistes luxatus*). The outlet of Upper Klamath Lake forms the headwaters of the Klamath River, the flow of which supports threatened anadromous fish populations and is important to tribal communities, anglers, and recreationalists. Managing the lake to meet all of these demands requires a quantitative understanding of the lake's hydrologic budget. Evapotranspiration from the lake and surrounding wetlands is a substantial component of the hydrologic budget of the lake.

Although hydrologic budgets have been developed for Upper Klamath Lake (Hubbard, 1970; Kann and Walker, 1999), estimates of evapotranspiration (ET) from the approximately 27,000 hectares (ha) of open water and 7,000 ha of wetlands surrounding the lake were obtained from pan-evaporation measurements and models—and have been a major source of uncertainty. Uncertainty in ET from Upper Klamath Lake and the surrounding wetlands propagates as uncertainty into other components of the budget including groundwater inflow; this uncertainty is problematic for lake management, water-supply forecasting, and management and restoration of drained wetlands.

Year-to-year operation of the Klamath Reclamation Project (which includes lake-level management) varies depending on statistically based water-supply forecasts made by the Natural Resources Conservation Service. One source of forecast uncertainty is that the regression equations use net inflow to Upper Klamath Lake as the dependent variable. Net inflow is a lumped term that includes multiple components of the lake budget such as inflow from streams and springs, ET, and precipitation. Forecasting equations would likely be improved if the dependent variable more directly represented the physical response of the watershed to measured conditions in the basin (precipitation, snow water content, and so forth) as would be the case if ET from the lake and surrounding wetlands could be independently determined.

In addition to the 7,000 ha of wetlands still surrounding the lake, roughly 12,600 ha of wetlands was separated from the lake by levees, drained, and put into cultivation over the past several decades. Large tracts of drained wetland area were purchased by the Federal Government or private conservation organizations between the mid-1990s and late 2000s. Understanding the consequences of various management options for these drained wetlands will be improved with a better understanding of wetland *ET* and open-water evaporation.

In a review of work to quantify predevelopment water budgets in the upper Klamath Basin (Bureau of Reclamation, 2005), the National Research Council (NRC; 2007) identified *ET* estimates as a major source of uncertainty in hydrologic calculations, and improving *ET* estimates was a specific recommendation of the NRC report. The current study was conceived to reduce uncertainty in estimates of *ET* in and around Upper Klamath Lake, thereby reducing uncertainty in the lake water budget, and improving comparisons between *ET* from different types of land use in the area.

Purpose and Scope

The purposes of this report are to (1) summarize measurements of year-round evapotranspiration from the dominant vegetative communities in natural wetlands surrounding Upper Klamath Lake; (2) summarize measurements of open-water evaporation that are more comprehensive than measurements made in the past; (3) use the resulting measurements to better calibrate standard *ET* models for use in water supply planning and water resource management in the basin; and (4) compare *ET* from natural wetlands, open water, and various types of agricultural lands in the area. This report describes the study area, underlying theory, methods of investigation, and results. Measurements of wetland *ET* were made at two sites from May 1, 2008, through September 29, 2010, encompassing three growing seasons. Measurements of open-water evaporation were made at two sites during the warmer months of 2008 through 2010.

Previous Studies

Bidlake (2000) measured *ET* by the eddy covariance method and calculated an energy budget at one location in the wetland northwest of the lake for four periods ranging from 1.2 to 1.9 days in 1997. While the work by Bidlake improved understanding of *ET* from wetlands around Upper Klamath Lake, significant improvements in understanding could be made by making similar measurements continuously for multiple growing seasons in a variety of vegetative settings.

Bidlake (2002) also measured *ET* by the Bowen-ratio energy balance method from three fallowed agricultural fields in the Tule Lake National Wildlife Refuge, California, about 45 kilometers (km) southeast of Klamath Falls, Oregon. These sites were selected to typify *ET* rates that might result from water conservation measures applied to formerly irrigated agricultural land. Alternative management strategies can be evaluated by comparing wetland *ET* measured in the current study to fallowed cropland *ET* estimated by Bidlake (2002).

Janssen (2005) calculated open-water evaporation from Upper Klamath Lake for part of summer 2003 using the energy-budget method and compared the results with rates determined using the modified Penman method (Penman, 1956). Janssen's work improved understanding of open-water evaporation from the lake and showed that the modified Penman method overestimated evaporation, but longer-term measurements are needed.

Description of Study Area

The upper Klamath Basin is semiarid but spans a steep precipitation gradient extending from the Cascade Range, where mean annual precipitation is 1.7 meters (m; Crater Lake National Park, Oregon, 1971–2000), to the interior parts of the basin where mean annual precipitation can be as low as 0.291 m (Tulelake, California, 1971–2000); all climate data in this section are from Western Regional Climate Center (2012). Mean annual precipitation (1971–2000) at Klamath Falls at the southern end of Upper Klamath Lake is 0.365 m. Of the total annual precipitation at Klamath Falls, 54 percent falls (mostly as snow) during the winter season (November through February), while only 15 percent falls during summer months (June through September). The remaining 31 percent falls during the transition months of October and March through May. Klamath Falls experiences cold winters with average December minimum and maximum temperatures of -5.6 degrees Celsius (°C) and 4.2°C, respectively, and warm summers with average July minimum and maximum temperatures of 10.9°C and 29.7°C, respectively (1971–2000).

Upper Klamath Lake (fig. 1) is the source of the Klamath River, and the lake encompasses approximately 270 square kilometers (km^2) with an average depth of roughly 2.8 m at full pool. The lake occupies a fault-bounded structural basin immediately east of the Cascade Range volcanic arc. Principal tributaries to the lake include the Williamson River, which accounts for about half the inflow, the Wood River, Sevenmile Canal, and several other smaller canals and streams originating from the Cascade Range (Hubbard, 1970). Most of the lake is just a few meters deep except for a trench that is as much as 15 m deep that runs along the western margin. Although Upper Klamath Lake is a natural water body, the stage of the lake has been increased and controlled artificially since 1921 by Link River Dam, a control structure built to facilitate management of the lake for irrigation. Total surface inflow to the lake is not routinely measured, but the mean annual flow of the Williamson River from 1971 to 2000 (measured below its confluence with the Sprague River, USGS station 11502500), which accounts for roughly half the total inflow, averaged about 31 cubic meters per second (m^3/s; 1,100 cubic feet per second (ft^3/s)). Outflow from the lake is principally to the Link

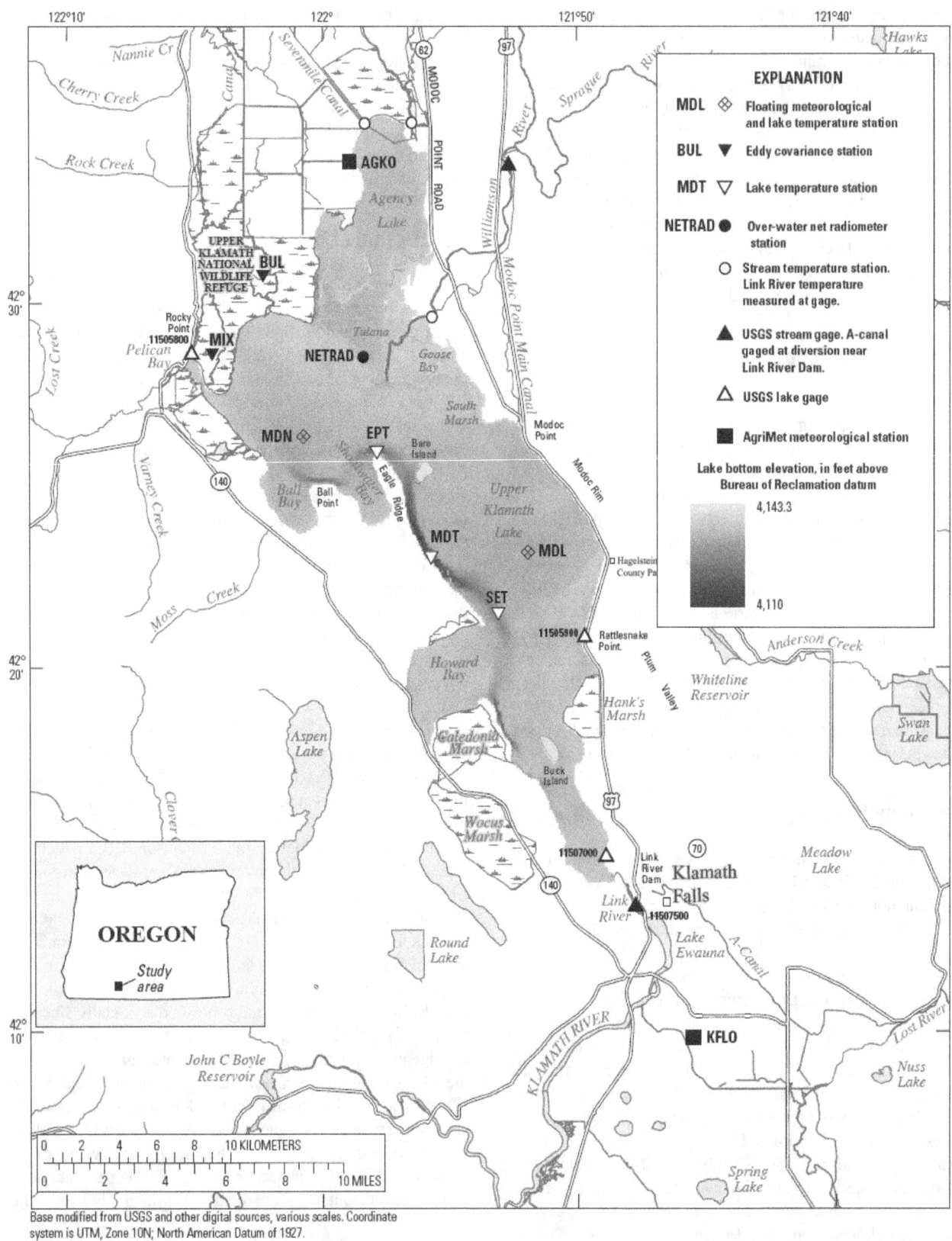

Figure 1. Upper Klamath Lake, surrounding area, and data collection locations.

River (which becomes the Klamath River downstream) and to the A Canal which delivers irrigation water to the east and south. Link River discharge averaged 38.57 m³/s (1,362 ft³/s) from 1971 to 2000 (USGS station 11507500), and annual diversions to the A Canal averaged 7.45 m³/s (263 ft³/s) during the same period (Jason Cameron, Bureau of Reclamation, written commun.). A third component of surface outflow is the Keno Canal. Reliable measurements are no longer available for this canal, but discharge averaged 5.32 m³/s (188 ft³/s) during the period from 1967 through 1983.

Upper Klamath Lake was originally surrounded by roughly 196 km² of wetlands. Between 1889 and 1971 about 126 km² was diked off and drained for agricultural use, whereas about 70 km² of natural wetlands remain (Snyder and Morace, 1997; Dan Snyder, USGS, written commun., 2011). The Upper Klamath Lake National Wildlife Refuge (NWR) is the largest natural wetland (58 km²) adjacent to the lake. The Nature Conservancy is in the process of restoring about 28 km² of previously drained wetlands on the Williamson River delta by removing or breaching dikes. Draining wetlands around the lake, however, has caused considerable oxidation and compaction of the thick peat soils, resulting in subsidence of the land surface by as much as 4 m (Snyder and Morace, 1997). As a consequence, water depths are too deep in most areas for reestablishment of wetland vegetation and the areas become open water when previously drained wetlands are reflooded. It is not known how much time will be required for reestablishment of wetland vegetation.

Upper Klamath Lake has been historically eutrophic but is now considered hypereutrophic. For the past several decades, the lake has experienced annual blooms of a near-monoculture of the blue-green algae *Aphanizomenon flos-aquae* which results in toxic conditions for fish (Snyder and Morace, 1997; Wood and others, 2006). During blooms, the algae congregate in the photic zone within a few centimeters of the lake surface, possibly affecting the albedo (ratio of reflected to incident light) and hence energy budget of the lake.

Study Sites

Two *ET* monitoring sites were chosen within the Upper Klamath National Wildlife Refuge (NWR) wetland area and two sites were chosen to monitor open-water evaporation (fig. 1). At full pool, the ratio of open water to wetland area is about 4:1. Despite this partitioning, two wetland sites were selected to address questions relating to differences in *ET* from different wetland communities and how these rates compare to *ET* from land historically drained to support irrigated agricultural crops, formerly irrigated land allowed to go fallow, and land recently returned from crop production to wetland status through breaching of dikes.

Bulrush and Mixed Vegetation Sites

The 5,800-ha Upper Klamath NWR is dominated by extensive wetlands and forms the northwest margin of Upper Klamath Lake, north and south of Pelican Bay (fig. 1), at an altitude of about 1,262 m. Lake level is controlled by the Link River Dam at the outlet and fluctuates about 1.3 m annually, causing large areas of the refuge (including the two study sites) to be flooded from about midwinter to midsummer of each year. The wetland areas of the NWR are covered in dense vegetation, heavily dominated by hard-stem bulrush (formerly *Scirpus acutus*, now *Schoenoplectus acutus*), with smaller amounts of cattail (*Typha latifolia*) and wocus (*Nuphar polysepalum*), and trace amounts of other vegetation. To typify *ET* from the most common species, one site was selected in a stand of almost exclusively bulrush, and another site was chosen within a patchwork of about 70 percent bulrush, 15 percent cattail, and 15 percent wocus. The sites were mapped on a visible light satellite image of the Upper Klamath NWR taken on August 28, 2011, and color patterns associated with each site were described. The bulrush site is located in an extensive, uniform, sepia-colored area, and the mixed vegetation site (hereafter called the mixed site) is located in a patchwork of sepia, light green, and dark green areas. Based on viewing the image of the whole NWR, we estimate that the bulrush site is typical of about 70 percent of the wetland, and the mixed site is representative of about 30 percent. Other objectives for site selection were to provide adequate fetch (extent of uniform vegetative cover around the sensors) for the *ET* measurements and to minimize the effort needed to access the sites during maintenance visits. These two objectives were best balanced by navigating along existing streams through the refuge far into the wetland, then along lateral, open, narrow, channels away from the streams, and finally out 30 to 50 m through the vegetation perpendicular to the channels. This strategy exploited the relatively easy travel along the streams and channels to penetrate deep into the wetlands and still kept the effects of the streams and channels on the *ET* measurements negligible. The two sites are located near the centers of the two largest lobes of wetland in the NWR (fig. 1). The bulrush site location is 42°30′48.88″ N., 122°2′4.89″ W., and the mixed site location is 42°28′36.80″ N., 122°4′6.05″ W. The bulrush and mixed site altitudes are 1,261.9 m (4,140.0 feet (ft)) and 1,262.0 m (4,140.5 ft) above mean sea level, respectively.

Bulrush, cattail, and wocus are phreatophytes, meaning their roots can remain flooded for extended periods without stress or damage caused by anoxia. In contrast, non-phreatophytic vegetation requires some degree of air penetration in the root zone to supply oxygen for root respiration. Consequently, the species at the study sites can thrive and transpire during the roughly 6-month period each year when lake water floods the local land surface, sometimes by as much as about 1 m. Bulrush and cattail are emergent vegetation, typically sending stalks and leaves to 2 to 3 m above the water surface. Their growth forms are similar, both having a central stalk and narrow, elongate leaves, all arranged nearly vertically when alive and vigorous. During winter dormancy, both plants cease photosynthesis, turn brown and brittle, and usually lodge over in response to snow and wind loading. At the study sites, the current year's fallen plants merge with the existing dead layers from past years, forming a loose network of branching debris, which persists year-round, averaging about 0.8 m in depth. Unusually high winds also caused patchy areas of live-vegetation blow-down, primarily at the mixed site during the 2009 growing season.

Wocus is a pond lily, classified as floating rather than emergent vegetation. Its roughly circular leaves are about 0.2 to 0.3 m in diameter and often rest on the water surface, distributed evenly, nearly covering the surface. At the study site, the plant sprouts and the leaves surface in the spring during the rising water level, remain at the water surface through midsummer, and become slightly aerial when water level recedes sufficiently in late summer. The leaves senesce and become part of the substrate as the water level approaches land surface. In early spring, areas occupied by wocus appear to be open water until the leaves surface.

Soils at the wetland study sites consist of a mat of roots and decaying plant matter from previous years, roughly 0.3 m thick, underlain by a saturated clay-silt-peat mixture sometimes called gyttja (Hansen, 1959; Snyder and Morace, 1997). The gyttja is basically a high-viscosity fluid just beneath the root mat, and it gradually grades into an elastic solid with depth. A small (about 5-millimeter (mm)) diameter rod pushed into the land surface encounters moderate resistance through the root mat and then penetrates easily below that point. Sizable impacts delivered to the land surface propagate laterally through the root layer as a wave, indicating the root layer is a somewhat tough skin, floating on the denser, more fluid gyttja below. This lithology posed unique challenges for working in the wetland and deploying the sensor stations, as discussed in Eddy-Covariance Sensors and Data Collection. The soil/root mat surface is hummocky, with a relief of about ± 5 centimeters (cm) over small horizontal scales (0.5 m), but the mat surface is extremely flat and level at horizontal scales larger than about 5 m.

Open-Water Sites

Open-water evaporation was measured using the Bowen-ratio energy balance (BREB) method (Anderson, 1954), with specialized sensors deployed at two open-water locations near the middle of Upper Klamath Lake, three open-water locations near the deep trench along the western margin of the lake, at the mouths of the Williamson and Wood Rivers (the two major inlets to the lake), the mouth of Sevenmile Canal, and at the beginning of the Link River and A Canal (fig. 1). In addition, lake stage was determined using existing sensors operated by the USGS located at Rocky Point, at Rattlesnake Point, and near the Link River Dam. Surface-to-air temperature and vapor-pressure differences, and water temperatures at various depths were measured at roughly the middle of the two largest lobes of the lake, at sites labeled MDL and MDN (fig. 1). These are previously established buoy locations used in other earlier and ongoing studies (Wood and others, 2006), where water depth is about 3 m. Minimum fetch to shore at the buoy sites is about 2.4 km, ensuring that the temperature and vapor-pressure differences were fully equilibrated to the lake surface (Stannard, 1997; Stannard and others, 2004).

The four components of net radiation were measured at a shallow open-water location near the shore at the mouth of the Williamson River (fig. 1). A four-component net radiometer was deployed at this site to investigate the effects of the high concentration of suspended algae on reflected solar (short-wave) radiation from the lake. In many lake evaporation studies, incoming radiation is measured on shore, and reflected solar radiation is modeled, assuming optical properties of clear water (Koberg, 1964; Sturrock and others, 1992). The known proliferation of algal blooms in Upper Klamath Lake suggested that a separate measurement of reflected solar radiation was warranted. The net radiometer site was chosen to typify the optical properties of water in the deeper parts of the lake but still provide water depths shallow enough to deploy a tripod on the lake bottom to support the sensor. Deployment from a raft was undesirable due to the inability to level the sensor reliably. Water depth at the site generally ranged from several centimeters to about 1 m, which was sufficient to behave optically like deeper midlake water during the bulk of the algal bloom periods.

Computation of open-water evaporation using the BREB method requires measurement or estimation of heat advected to or from the water body caused by inflows and outflows of surface water and groundwater. Therefore, water temperature and stage were measured at the two major surface-water inflows (the mouths of the Williamson and Wood Rivers) and at the Link River and A Canal outflows (fig. 1). Other surface water inputs were much smaller and were, therefore, not instrumented. The net heat exchange with groundwater also was considered to be minor (Janssen, 2005), and no attempt was made to measure it.

Methods of Study

Evapotranspiration (*ET*) of water is an important component of the hydrologic cycle. Globally, *ET* is equal to precipitation (*P*), and over land, *ET* is about 62 percent of *P* (Brutsaert, 1982, table 1.1), larger than any other terrestrial hydrologic component. The importance of *ET* is underscored by recognizing that once water evaporates into the atmosphere, it is no longer available for human consumption until it returns to the surface as *P*, which led to the adoption of the term consumptive use as a synonym for *ET*. Although the ratio *ET:P* decreases in humid climates (generally reducing the importance of *ET* there), *ET:P* approaches unity as aridity increases, making *ET* extremely important in semiarid and arid parts of the world.

The word evapotranspiration was coined in the 1930s to combine the processes of evaporation and transpiration into a single term (Brutsaert, 1982). Whereas evaporation is the phase change of liquid water to water vapor, transpiration is evaporation that specifically occurs within the substomatal cavities of vegetation. Plants withdraw soil water and nutrients from the root zone and transport them up through the plant stems to the leaves, where the nutrients are used by plant tissues and the water evaporates through microscopic holes in the leaves called stomata. Current usage of *ET* refers to the sum of evaporation from open water, wet plant and mineral surfaces (such as shortly after rainfall), sublimation from snow and ice, and transpiration. In this report, *ET* refers to water loss from vegetated sites to the atmosphere, and evaporation (*E*) is reserved for water loss from extensive open-water sites to the atmosphere. Both *ET* and *E* can be expressed in a variety of units. In this report, daily and biweekly totals are expressed in millimeters per day (where millimeters is depth of liquid water per unit area of land surface); these basic units are aggregated into meters per year to describe annual evaporative losses.

Evaporation of water consumes a relatively large amount of energy, known as the latent heat of vaporization of water, *L*. Consequently, *ET* is a substantial component of the energy balance of the Earth's surface, as well as the hydrologic balance. As liquid water vaporizes at the surface, it carries energy into the atmosphere, which is released at some later time when the vapor re-condenses into liquid.

The rate of energy used to sustain a given *ET* rate is known as the latent-heat flux (*LE*) and is equal to:

$$LE\left[\frac{W}{m^2}\right] = ET\left[\frac{mm}{d}\right] \times \frac{1}{86400}\left[\frac{d}{s}\right] \times \frac{1}{1000}\left[\frac{m}{mm}\right] \times \rho_w\left[\frac{kg}{m^3}\right] \times L\left[\frac{J}{kg}\right] \quad (1a)$$

where

ρ_w is the density of water,
L is the latent heat of vaporization, and
W   is joule per second.

Both ρ_w and *L* are very weak functions of temperature, and they are roughly constant over the range of temperatures encountered in this study. Therefore equation 1a can be rewritten approximately as:

$$ET\left[\frac{mm}{d}\right] = 0.0351 \times LE\left[\frac{W}{m^2}\right] \quad (1b)$$

Energy Balance of a Vegetated or Open-Water Surface

Virtually all of the energy arriving at the Earth's surface originates from the sun (Kiehl and Trenberth, 1997). On a global, annual average basis, solar (short-wave) radiation arriving at the top of the atmosphere is 342 watts per square meter (W m^{-2}), and is reduced to 198 W m^{-2} at the Earth's surface. On average, 168 W m^{-2} of the solar radiation arriving at the surface is absorbed, and 30 W m^{-2} is reflected, giving an average reflectivity (a, also known as albedo) of 0.15. Some of the solar radiation arriving at the top of the atmosphere is reflected to space (77 W m^{-2}) or absorbed by the atmosphere (67 W m^{-2}). Atmospheric absorption occurs primarily by clouds, water vapor, dust, trace gases, and aerosols. The absorbed energy heats the atmosphere, which in turn emits long-wave (infrared) radiation, some of which arrives at the surface. The surface either reflects or absorbs this long-wave radiation, and also emits its own long-wave radiation. The processes of reflection, absorption, emission, and transmission of long-wave radiation between the surface and atmosphere are complex, but on average, the net long-wave exchange is 66 W m^{-2} from the surface upward to the atmosphere.

At the Earth's surface, the algebraic sum of incoming solar radiation (positive), reflected solar radiation (negative), incoming long-wave radiation (positive), and outgoing long-wave radiation (negative) constitutes net radiation, R_n. On average, R_n is (198 minus 66, or) 102 W m^{-2} (Kiehl and Trenberth, 1997) and drives all of the other (nonradiative) energy exchanges at the surface. A schematic of the typical daytime energy balance of a vegetated surface is shown in fig. 2A. The surface typically warms during the day in response to the input of energy from R_n. Some of that energy moves into the subsurface by conduction, known as the soil heat flux, *G*.

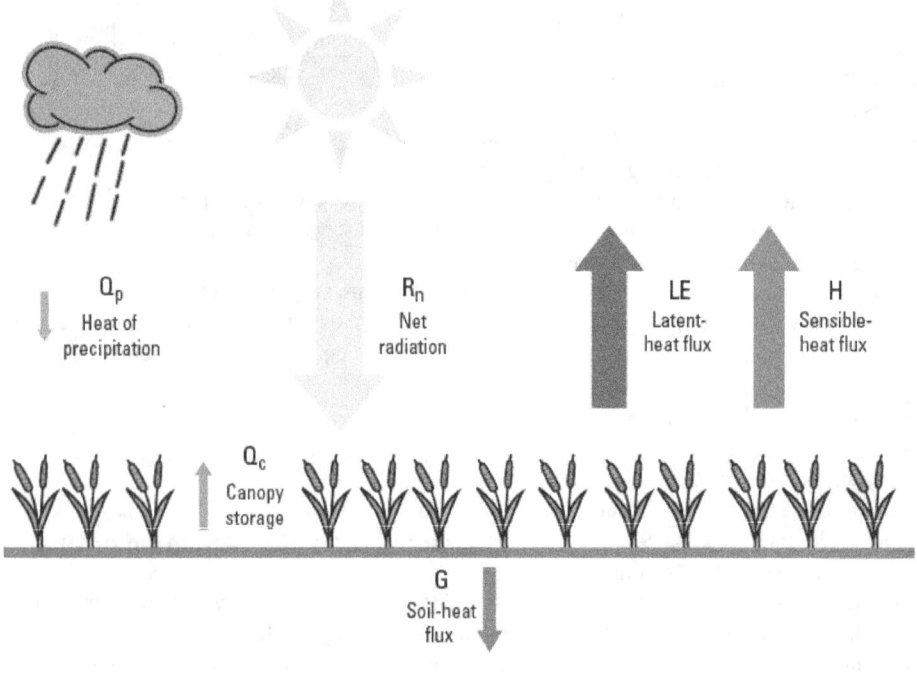

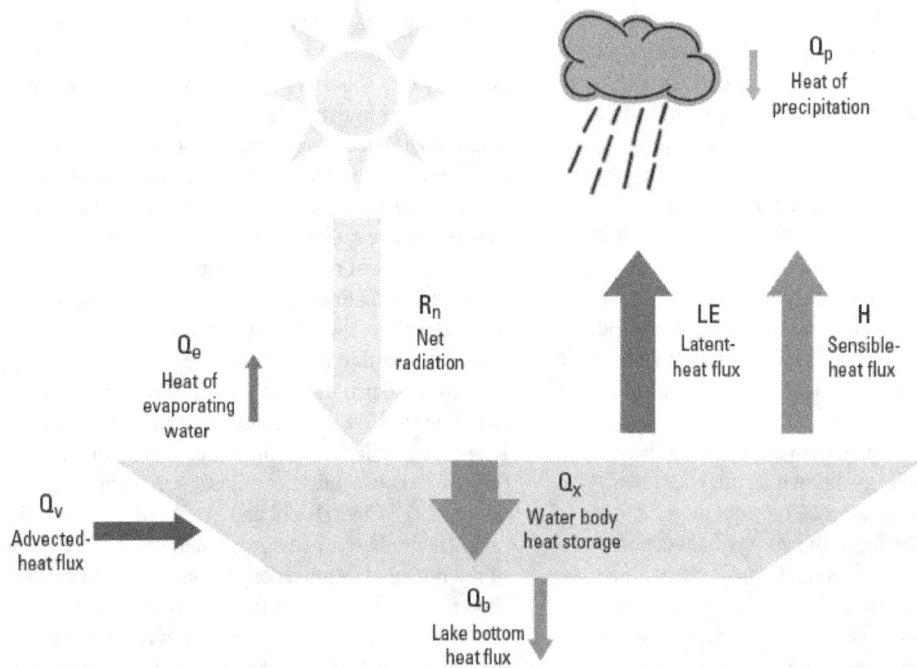

Figure 2. Energy balance of *A*, a vegetated surface and *B*, a water body.

The portion that warms the vegetation biomass is canopy storage, Q_c. On most days, absorbed radiation warms the vegetation and soil surfaces above the temperature of the overlying air, giving rise to turbulent transfer of heat into the air, known as sensible-heat flux, H. Water evaporating from the surface carries with it the latent-heat flux, LE, another turbulent flux from the surface to the air. Global annual average values of H and LE are 24 and 78 W m^{-2}, respectively. Finally, the excess heat content of precipitation above the surface temperature produces a typically small flux, the heat of precipitation, Q_p. Adopting the sign convention shown in figure 2A, the surface energy-balance equation can be written as:

$$R_n - G - Q_c + Q_p = H + LE \qquad (2)$$

The left side of equation 2 is called the available energy (AE), and the right side is the turbulent flux (TF). In many settings, Q_c and Q_p are minor terms, leaving R_n, G, H and LE as the four main components of the energy balance. In this study, Q_c and Q_p were determined to be negligible and are not included in the calculations.

At night, solar radiation terms become zero and outgoing long-wave radiation typically exceeds incoming radiation, causing R_n to become negative. Usually, the surface cools faster than the overlying air due to this long-wave emission, creating a temperature inversion in the air, leading to a sign change in H and a downward flux of sensible heat. Typically, the surface is also cooler than the subsurface soils, and G similarly changes sign. Outgoing night-time G is comparable to incoming daytime G, and long-term values of G approach zero. Most plants close stomata at night, shutting off transpiration, and due to the lack of available energy and colder temperatures, evaporation from soil also approaches zero, resulting in a surface LE that approaches zero at night. If the surface cools below the dew-point, water vapor in the air deposits as dew (or frost when the surface is below freezing) and LE becomes slightly negative.

The dynamic changes in the major fluxes between day and night result in typical 24-hour time series of flux resembling sinusoidal waves, with maxima near solar noon and minima near midnight. Daily average amplitudes and means are greatest in summer and least in winter. In addition, substantial variability about the average patterns occurs, caused by changes in the amount of cloud cover, green vegetation, water availability, wind speed, and atmospheric properties.

The surface energy balance shown in figure 2A is a one-dimensional, vertical model, written for a specific, uniform surface of interest. Energy balances can vary substantially from one surface type to another. Because the turbulent fluxes are measured at some height above the plant canopy, the assumption is made that they are equal to the values at the surface, corrected for any change in heat or vapor storage between the canopy and the sensor. This assumption is valid if the horizontal extent of the uniform surface of interest in the upwind direction (the fetch) is sufficiently large that the air passing by the turbulent flux sensors has equilibrated to the surface of interest. Consequently, turbulent flux sensors typically are deployed in the midst of a large expanse of a uniform surface type. Tests are available to evaluate the adequacy of the fetch (Schuepp and others, 1990; Stannard, 1997) and are used in this study.

Compared to the energy balance of a land surface, the balance of an open-water surface is complicated by the subsurface transfer of heat in three dimensions as the bulk liquid water mixes in response to wind, density gradients, inflows, and outflows. In addition, solar radiation typically penetrates the water column to some depth, further complicating the subsurface energy exchange. To avoid the problems of measuring the near subsurface energy input, the entire water body is taken as a control volume, and energy fluxes into and out of that volume (rather than a surface) are conceptualized and measured (fig. 2B). The fluxes of R_n, H, LE, and Q_p are analogous to those occurring over land. The flux of heat energy exchanged between the surface and subsurface (the equivalent of G on land) is recast as a change in heat stored in the water body, Q_x, measured by changes in temperature profiles through the water column (Anderson, 1954; Sturrock and others, 1992). If the water body is relatively shallow, heat exchange also may occur between the water and the lake bottom sediments, designated as Q_b. Water bodies also are often connected to streams and groundwater, which can add or remove heat to or from the water body through advection, designated as Q_v. Finally, heat can be added or removed to or from the water body by the evaporating water, depending on whether that water is cooler or warmer, respectively, than the mean water-body temperature. Designated as Q_e, this small flux also occurs over land but is seldom evaluated there because the temperature of the evaporating water usually is not known. The resulting energy balance of a water body can be written as (Anderson, 1954):

$$R_n - Q_x - Q_b + Q_v - Q_e + Q_p = H + LE \qquad (3)$$

The diurnal behavior of energy-balance fluxes is quite different over water than over land, primarily because (1) solar radiation partially penetrates the surface, (2) the thermal inertia of water is much greater than that of land, and (3) water availability at the water surface is always 100 percent. These factors combine to create a daytime balance where heating of the water surface from R_n occurs more slowly than heating of the land (or vegetation) surface, moderating surface temperature compared to temperature over land. Usually the overlying air heats more quickly than the water surface, resulting in a temperature inversion and a downward (negative) H during the day. Rather than being partially converted into a positive H as over land, R_n (aided by a downward H) drives a large value of Q_x by storing heat at depth in the water column, and the residual energy is partitioned into LE. At night, the air cools faster than the water surface, and H becomes positive. Even though R_n is negative, heat comes back out of storage from the water column ($Q_x > -R_n$), driving both a positive LE and a positive H. Daytime Q_x over water is a large fraction of R_n, whereas $G + Q_c$ usually is a small fraction of R_n over a densely vegetated canopy on land. The 24-hour wave-like behavior of H over water is about 12 hours out of phase with H over land, and magnitudes are much smaller. The 24-hour behavior of LE over water is less wave-like, tending toward a more constant (positive) value fluctuating about the 24-hour mean.

The two wetland sites of the current study exhibit energy-balance characteristics of both the vegetated land surface and the water body described in the previous several paragraphs. Between midsummer and midwinter, standing water is not present at the sites, and the surface behaves purely as a vegetated land surface. From midwinter to midsummer, water depth at the sites typically increases to as much as about 0.8 m, is maintained at that level for about 2 months, and then recedes back to zero. During this period, diurnal energy flux patterns are intermediate between those of a vegetated land surface and an open-water body, depending on vegetation status and water depth. The energy balance of a wetland surface with standing water contains elements of equations 2 and 3, and can be written:

$$R_n - Q_x - G = H + LE \tag{4}$$

At times when standing water is present, the dense vegetation canopy effectively inhibits significant lateral exchange of energy in the water column, and therefore Q_v is assumed to be zero. At times when standing water is not present, Q_x and Q_v are both zero.

Eddy-Covariance Measurements of Latent- and Sensible-Heat Flux

Turbulent exchange of energy and mass in the lowest layer of the atmosphere, the atmospheric boundary layer (ABL), occurs through eddy diffusion (Brutsaert, 1982, p. 52–56). Above a flat, level surface, the mean air movement is commonly recognized as being horizontal and is quantified as wind speed. However, superimposed on the mean wind are random turbulent upward and downward motions, resulting in instantaneous wind vectors with non-zero vertical components. The interplay between air viscosity, geostrophic wind (wind at upper levels of the atmosphere caused by regional pressure differences), surface heating, and surface roughness creates an ABL where air motions tend to aggregate into blobs or eddies which span a wide range of sizes, from centimeters to kilometers. Within an eddy, movements are highly correlated and can be quite distinct from those in nearby eddies. Similarly, concentrations of admixtures (for example, the heat, water vapor, or carbon dioxide content) may vary from eddy to eddy. Vertical transport of eddy admixtures results from the vertical components of eddy motions, a process similar to molecular diffusion but on a much larger scale. As an eddy moves upward or downward, it carries the admixtures along with it, giving rise to vertical turbulent flux. Above an extensive uniform surface, the net vertical transport of an admixture is equal to the algebraic sum of the transport contributions from all eddies (or a statistically significant sample thereof) passing over the surface. Thus, over an evaporating surface, upward-moving eddies contain, on average, more water vapor than downward-moving eddies. If certain criteria are met, the flux of an admixture can be measured through a high-speed bookkeeping of vertical eddy motion and admixture concentration, known as eddy covariance (EC). In the case of water vapor, the EC equation takes the form (Dyer, 1961):

$$ET_m = 86.4 \cdot \overline{w' \rho_v'} \tag{5}$$

where

- ET_m is measured evapotranspiration, in millimeters per day;
- w is vertical component of wind speed, in meters per second (upward is positive);
- ρ_v is vapor density, in grams per cubic meter; primes denote deviations from mean values; overbar denotes mean value; and the factor 86.4 converts grams per square meter per second into millimeters per day.

The quantity $\overline{w'\rho_v'}$ is the covariance of w and ρ_v. Similarly, the equation for sensible heat is:

$$H_m = \rho_a c_p \times \overline{w'T_a'} \tag{6}$$

where

> H_m is measured sensible heat flux, in watts per square meter;
>
> ρ_a is air density, in kilograms per cubic meter;
>
> c_p is specific heat of air at constant pressure, in joules per kilogram per degree Celsius; and
>
> T_a is air temperature, in degrees Celsius.

Typically, means are computed over 20 to 60-minute periods, and deviations from means are measured 5 to 20 times per second.

In addition to turbulent flux of heat and mass between the surface and atmosphere, a turbulent flux of horizontal momentum also occurs from the atmosphere to the surface. Horizontal wind speed is zero at the surface and increases with height above the surface—a wind-speed profile that indicates greater momentum at height, decreasing to zero momentum at the surface. This gradient implies a flux of horizontal momentum from the upper atmosphere downward to the surface, also recognized as a drag force exerted on the surface by the air. The transfer of momentum downward occurs through eddy diffusion and can be expressed as:

$$\tau_m = -\rho_a \times \overline{w'u'} \tag{7}$$

where

> τ_m is measured momentum flux or drag per unit area of land surface, in kilograms per meter per second squared; and
>
> u is horizontal wind speed, in meters per second.

Usually, τ_m is measured concurrently with ET_m and H_m because it is needed to calculate corrections to those fluxes.

The Energy-Balance Closure Problem

The surface energy balance illustrated in figure 2A and quantified in equation 2 accounts for all significant energy fluxes to and from a surface that are currently recognized (Wilson and others, 2002). Commercially available instrumentation provides measurements of each term in equation 2 separately. If all measurements are accurate and all significant fluxes appear in equation 2, then the left side of the equation, the available energy (AE), should be equal to the right side, the turbulent flux (TF), within the limits of measurement accuracy. This concept is commonly called energy-balance closure. Graphical and statistical comparison of AE with TF frequently is used to identify problems

associated with sensor inaccuracies and malfunctions, sensor deployment, insufficient fetch, or complex terrain. Due to random components of some errors, measurement accuracy usually increases as the accumulated time interval of continuous measurements increases. In addition, some errors may be systematically linked to time of day or year, but are largely self-compensating when the measurement interval is a full day, a number of full days, or a full year. Consequently, energy-balance closure usually is evaluated over longer time periods, preferably a year or multiple years.

Comparisons of long-term AE with TF from many studies display a chronic mismatch between the two. The ratio of TF to AE generally ranges from about 0.6 to 1.0, it most frequently is around 0.7 to 0.8 (Twine and others, 2000; Wilson and others, 2002; Foken, 2008). In the past, undermeasurement of TF (Dugas and others, 1991; Twine and others, 2000; Gash and Dolman, 2003) or overmeasurement of R_n (Campbell, 1980; Halldin and Lindroth, 1992; Bossong and others, 2003), both presumably caused by instrumental performance or deployment, have been suggested as reasons for the mismatch. Soil-heat flux and storage terms are not suspect because long-term measured values approach zero, consistent with principles of environmental physics. Recently, Foken (2008) argued that current instrumentation and field practices are now sufficiently accurate that they can no longer account for the lack of closure. Rather, Foken proposed that land-surface scales of heterogeneity on the order of 1 to 10 km may generate large-scale eddies that contribute to turbulent flux in a fashion that is unlikely or impossible to be fully measured using the EC method. Foken argues that the time of passage of these very large eddies may be too long for typical averaging periods (which are 30 minutes to 1 hour; longer averaging periods violate the stationarity principle) to capture. Further, he suggests that the large eddies cluster near substantial changes in land-surface type (the very locations that are avoided as measurement sites if the goal is to obtain a clear, unambiguous flux signal from a surface of interest), inducing undetected horizontal flux advection over the central areas of homogenous land-surface patches, where sensors typically are deployed. Large-scale (~5 km) scintillometer measurements of turbulent flux and large eddy simulation modeling results support these ideas, both indicating adequate energy-balance closure when applied to horizontal scales of several kilometers encompassing changes in land-surface type (Foken, 2008). Although a proven remedy for ordinary EC data collection is not offered, Foken suggests as a first guess closing the energy balance by adjusting both H and LE, maintaining the ratio between them ($H/LE = \beta$, the Bowen ratio). This method has also been used commonly in the past (Barr and others, 1998; Blanken and others, 1998; Twine and others, 2000) when energy-balance closure was not obtained. Although Foken's (2008) ideas are promising, the scientific community is still undecided on the answer to the energy-balance closure problem.

In this report, measured H and LE are corrected using the energy-balance closure principle. A single value of the energy-balance ratio is computed as:

$$EBR = \frac{\sum H_m + \sum LE_m}{\sum R_n - \sum Q_x - \sum G} \tag{8}$$

where

EBR		is the mean energy-balance ratio (unitless) during the study period (May 1, 2008 to September 29, 2010);
$\sum$		indicates the sum of all measured values during the study period, using only 30-minute periods when all energy-balance sensors were operating correctly; and
H_m and LE_m		refer to the measured values of H and LE, uncorrected for energy-balance closure.

First, gap-filling and standard corrections (described in Eddy-Covariance Data Processing) are applied to the 30-minute H and LE data. Then the resulting H_m and LE_m values are divided by EBR to obtain final 30-minute values that produce long-term energy-balance closure. The

use of a single, mean value of EBR, rather than multiple values calculated over shorter periods, is discussed in Energy-Balance Closure and Trends. The EBR values at both EC sites are reported in Energy-Balance Closure and Trends, allowing the reader to compute H_m and LE_m (or ET_m) if desired. Applying this EBR correction to the measured wetland EC fluxes facilitates direct comparison with measured open-water evaporation, which is based on the concept of full energy-balance closure.

Eddy-Covariance Sensors and Data Collection

The EC method requires fast-response sensors to adequately measure the rapid fluctuations in wind-speed components, air temperature, and vapor density caused by movement of the smallest eddies that contribute to the turbulent flux of momentum, heat, and water vapor. Two specialized sensors developed for eddy-covariance measurements by Campbell Scientific were used in the present study. A 3-axis sonic anemometer (Model CSAT3) measured wind vectors along three measurement paths, all at 60 degrees (°) from vertical, and at 120° from each other in the horizontal plane (fig. 3). The sensor was leveled in the field by means of a sensitive bubble level. Software in the CSAT3 converted the measured vectors into 3 orthogonal vectors aligned with

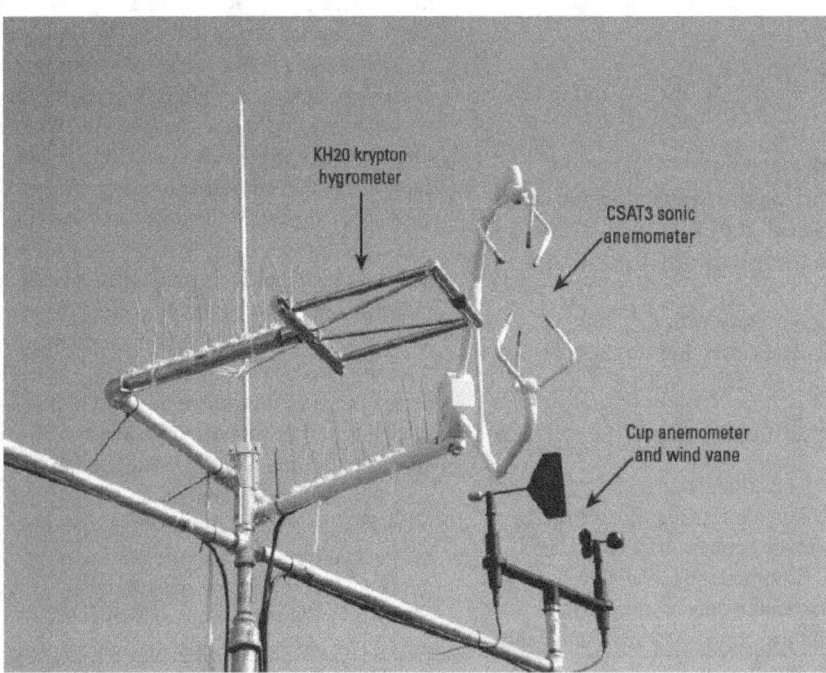

Figure 3.   Eddy-covariance sensors, cup anemometer, and wind vane.

the sensor's major axes (one vertical and two horizontal). The CSAT3 measured the wind vectors by sending ultrasonic signals between transducer pairs (fig. 3), measuring the travel time, and using the Doppler effect to compute the wind speed. The two orthogonal horizontal vectors were used to compute the resultant horizontal vector (u) appearing in equation 7. The CSAT3 also measured sonic air temperature, T_s, using the dependence of the speed of sound on temperature. This sonic air temperature is slightly different than true air temperature, T_a, and requires a small correction

to obtain the $\overline{w'T_a'}$ covariance (eq. 6), described in Eddy-Covariance Data Processing. A krypton hygrometer (Model KH20) was deployed adjacent to the CSAT3 to sense vapor density fluctuations (fig. 3). The source tube emits a krypton radiation signal through a window, along the short (1 cm) measurement path, and through another window to the receiver tube. Because the radiation is absorbed by water vapor, the signal strength received is inversely proportional to the vapor density in the measurement path, providing a virtually instantaneous measure of ρ_v'.

The EC sensors were deployed from a galvanized steel tripod and were oriented to minimize interference upwind of the measuring paths during wind flow from the prevailing directions. The tripod was installed immediately adjacent to a platform structure that supported all of the other sensors and hardware and provided a working surface during installation and site visits (fig. 4). Isolation of the tripod from the platform prevented disturbing the EC sensors while researchers were moving about on the platform during site visits. The platform and tripod required special modifications to obtain adequate support from the spongy peat soil underlain by gyttja. Slow settling of the structures into the peat could affect sensor levels unacceptably, and penetration into the semiliquid gyttja below could topple the entire station. Large wooden feet made from 2 by 10-inch treated lumber were attached to the bottom of the tripod legs—and to the platform legs with additional sections of leg protruding below the feet to puncture the peat layer and anchor the platform (fig. 5). This design was sufficient to prevent any noticeable differential settling of the structures, thereby maintaining sensors level within acceptable limits between site visits. The EC sensors were set at heights of 3.61 m and 3.76 m above land surface at the bulrush and mixed sites, respectively. At both sites, the KH20 measurement path was located 10 cm from the midpoint of the CSAT3 measurement paths (fig. 3).

Figure 4. Completed eddy covariance, energy balance, and meteorological station at bulrush site on June 10, 2008. Photo courtesy of Brian Wagner, U.S. Geological Survey.

A   *B*

Figure 5.   Specialized feet to support *A,* energy balance and meteorological platform and *B,* eddy-covariance tripod, on spongy wetland surface. Platform shown just before installation into wetland surface.

The response times of the CSAT3 and KH20 sensors are nearly instantaneous, ensuring that rapid sampling of their outputs accurately tracks the random variations in wind, temperature, and vapor density. All 5 EC variables (three orthogonal wind vectors, sonic air temperature, and vapor density) were sampled at 10 hertz (Hz), using a Campbell Scientific CR1000 data logger. The 10 Hz data were temporarily stored in the data logger during each 30-minute period. These data were then used by the logger to compute means and standard deviations of each variable and all (10) of the possible covariances between variables (using block averaging) at the end of each period. In addition, a time stamp (year, month, day of month, hour, minute), preliminary flux values, wind direction, data-logger temperature, system battery voltage, and sensor diagnostics were recorded at the end of each period. Data were sent automatically via cell phone modem to the USGS Water Science Center in Portland, Oreg., daily. At each site, power was supplied from a single 40 W solar panel, with storage initially provided by a 100 ampere-hour deep cycle flooded battery (fig. 4). Subsequently a second battery was added at each site after night-time low-power problems occurred at the mixed site.

Eddy-Covariance Data Processing

Although the EC method is the most direct measurement of turbulent fluxes between the surface and atmosphere currently available, the apparent simplicity of equations 5 through 7 is slightly complicated by the need for minor corrections arising from limitations of the method and sensors. The following commonly used corrections were applied to EC data in the present study.

First, the use of w' in equations 5 through 7 is predicated on the assumption that mean wind flow past the sensors is horizontal (that is, $\bar{w} = 0$). Over surfaces with uneven topography, vegetation, or heating, 30-minute mean wind vectors may not be horizontal, and the 30-minute mean vertical component may change depending on wind speed, wind direction, and solar heating. Although land surface at both EC sites was very flat and level, disturbances to the vegetation occurred from station installation and from wind gusts that lodged patches of bulrush and cattail, creating depressions in the canopy in the vicinity of the stations. Coordinate rotation was, therefore, performed to correct the three covariances in equations 5 through 7 (Tanner and Thurtell, 1969). This correction requires computing second-order moments and may increase or decrease the calculated flux, depending on the vertical direction of the rotation. Rotations were made based on the 30-minute mean wind vector, rather than using the long-term planar fit method (Paw U and others, 2000), which is more appropriate when the land surface or the sonic anemometer is not level.

Second, above even a flat, level surface, a more subtle effect contributes to a nonzero $\bar{w}$ when $H \neq 0$, caused by the inverse relation between air temperature (T_a) and density (ρ_a) at constant pressure. For example, if $H > 0$, rising eddies are warmer and, therefore, less dense than sinking eddies. Because the vertical flux of dry air over a horizontal surface is zero, a small (upward) $\bar{w} > 0$ occurs to maintain conservation of mass (Webb and others, 1980), and the converse occurs ($\bar{w} < 0$) when $H < 0$. A correction to the measured $\overline{w'\rho_v'}$ (eq. 5) was made to remove the effect of $\bar{w} \neq 0$ following Webb and others (1980). This correction increases ET when $H > 0$ (usually daytime) and decreases ET when $H < 0$ (usually nighttime), generally producing a net increase in daily ET.

Third, a small correction to the measured $\overline{w'\rho_v'}$ covariance in equation 5 is needed because the krypton radiation emitted by the KH20 source tube is slightly attenuated by oxygen, and oxygen density is proportional to total air density. At constant pressure, warmer eddies are less dense, creating a small error in the measured $\overline{w'\rho_v'}$ when $H \neq 0$. Following Tanner and Greene (1989), a correction proportional to H was made to the measured $\overline{w'\rho_v'}$, to remove this artifact.

Fourth, the sonic air temperature, T_s, is slightly different than true air temperature, T_a, because T_s is affected by the water vapor content of the air. Following Schotanus and others (1983), a small correction proportional to LE was made to convert $\overline{w'T_s'}$ to $\overline{w'T_a'}$. Because the second, third, and fourth corrections are interdependent, these corrections were iterated until convergence was obtained.

Fifth, frequency response corrections (Massman, 2000) were made to compensate for the inability of the EC method to record flux contributions from the largest and smallest eddies under certain conditions. The largest eddies contributing to flux may require longer than 30 min to pass by the sensors in low wind speeds (less than about 0.5 m/s). However, averaging periods longer than 30 min begin to violate the stationarity principle and are generally not used. At the other extreme, contributions to flux from the smallest eddies (about 10 cm) may not be fully recorded due to sensor geometry such as path-length averaging and sensor separation. These corrections apply to H, LE, and τ and the corrections to each flux depend on the magnitudes of the other fluxes, so these corrections also were iterated until convergence was obtained.

The calibration factor of the KH20 varies slightly with ambient vapor density, ρ_v. The manufacturer supplies the calibration curve data points, and the slopes of least squares linear regression fits to the data in the high and low ρ_v ranges. The manufacturer suggests using one or the other slope as the calibration factor, depending on ambient ρ_v. This approach creates unnecessary discontinuities at the transition and small errors in computed ET. We improved upon the piecewise linear approach by fitting a cubic polynomial to the calibration data ($r^2 > 0.999$) and using a more exact calibration factor for each 30-minute period, equal to the slope of the cubic curve (a quadratic function of ρ_v).

Fully processed 30-minute ET and related data are aggregated into longer period averages and totals for presentation, seasonal analysis, comparison with other work, and computation of crop coefficients. Depending on the application, periods of 1 day, 2 weeks, a growing season, or 1 year are used as averaging periods.

Gap-Filling Missing or Bad Eddy-Covariance Data

The EC sensors can malfunction when their measurement paths or surfaces are obstructed by water droplets, water films, snow, ice, particulates, leaves, insects, webs, avian fecal matter, or other solids or liquids that interfere with the transmission or reception of the sensor signals. By far, the most common interference during this study occurred from precipitation, and it affected primarily the KH20 hygrometer. The CSAT3 sonic anemometer also was affected at times by the presence of liquid or solid water, but much less frequently than the KH20 hygrometer. Once the measurement paths and surfaces of either sensor are sufficiently free of water and other foreign matter, correct sensor operation and data collection resume.

Periods of invalid data were identified by graphing the four energy-balance fluxes, precipitation, air temperature, vapor density, the KH20 voltage signal, horizontal wind speed, and momentum flux for periods of a week and searching for anomalous patterns in H, LE, and τ. Anomalous patterns in H and LE (extended periods of zero flux, spikes, noise) were identified based on energy-balance principles. Periods of suspect data were then graphed on an expanded time scale (1 or 2 days per graph) to identify invalid data by 30-minute intervals. In almost all cases, reasons for invalid flux data were identified by inspection of the auxiliary data from the same period. For example, recorded precipitation, a sudden decrease in the KH20 voltage signal without a corresponding increase in vapor density, and relative humidity near 100 percent, were commonly noted during invalid flux data collection and strongly suggested that water interfered with one or both EC sensors.

Based on the close inspection of flux and auxiliary data, flags were added to data records to indicate: (1) invalid data, (2) which fluxes were affected, and (3) the appropriate gap-filling method to use. By inspection, it was found that the status of H and τ (operational or invalid) were identical, requiring only a single flag to indicate the status of both fluxes. Gap-filling was accomplished through linear interpolation or through energy-balance modeling. Linear interpolation was used when the gap was sufficiently short (≤ 2 hours) or the auxiliary data suggested that the missing flux varied approximately linearly between the last valid data point before the gap and the first valid data point after recovery. During the growing season, this second condition typically was restricted to periods of 8 hours or less because the diurnal cycle typically imposes a curvature in energy-balance fluxes during longer periods. During winter, periods as long as 18 hours were interpolated if the data indicated low sunlight (overcast), cold temperature, and precipitation—a winter storm. This practice is reasonable because with little diurnal variation in available energy, little resulting diurnal variation in turbulent flux occurs, creating extended periods when H and LE are small and relatively constant. Energy-balance modeling was required to gap-fill LE data only, because all gaps in H and τ were sufficiently short to gap-fill using linear interpolation. Energy-balance gap-filling of LE incorporated the concepts embodied in equations 4 and 8, and took the form:

$$LE_m = EBR(R_n - G - Q_x) - H_m \qquad (9)$$

This method estimates LE using the energy-balance closure principle (eq. 4), modified by the observation that the long-term ratio of measured turbulent flux to available energy (that is, the EBR) usually is less than one (eq. 8).

Periods of missing data occurred at the mixed vegetation site, caused by inadequate battery storage during June 14–25, 2008, and by cell-phone malfunction during July 30–August 12, 2009. The first incident was a series of 13 night-time gaps, separated by daytime periods of successful data collection when adequate power was supplied directly from the solar panel. The second gap was a 13.1-day period of no data collection. During the low battery power gaps, both EC and meteorological data were missing, whereas during the cell-phone gap, only EC data were missing. Because the two EC sites are of similar land cover and are separated by only 5.8 km, energy fluxes and most weather conditions are very highly correlated at the two locations. Periods of missing data at the mixed site were, therefore, gap-filled using simple linear regression against the equivalent data collected at the bulrush site. The linear relations were established using periods immediately before and after each gap, when data collection at both sites was successful. The periods before and after the gap were each about one-half the length of the gap, or longer. In the case of the recurring night-time gaps, round-the-clock data collected before the first occurrence of low power and after the problem was remedied were used to create robust regressions (exploiting the full diurnal range in flux), even though only night-time periods were gap-filled.

Meteorological, Energy-Balance, and Biophysical Measurements at Eddy-Covariance Sites

Meteorological and Energy-Balance Sensors and Data Collection

In addition to basic EC data collection, sensors were deployed at both EC sites to measure meteorological and energy-balance variables: (1) to help identify times when EC sensors were inoperative due to environmental moisture (rain, snow, ice) or other interference; (2) to gap-fill compromised EC data during these times; (3) to evaluate the degree of energy-balance closure; (4) to associate ET and energy-balance results with environmental conditions; and (5) to evaluate the fetch at the EC sites. Variables measured were net radiation, soil-heat flux, wind speed, wind direction, soil water content, rainfall, air temperature, relative humidity, and water temperature at various depths (when standing water was present). Sensor make and model, measurement height, scan interval, and summary statistics are presented in table 1.

The summary interval for all sensors except the water-temperature sensors was 30 min. The summary interval for the water temperature sensors was 15 min. The CNR1 net radiometer consisted of four separate sensors to measure incoming solar, reflected solar, incoming long-wave and outgoing long-wave radiation. The rainfall gage was a nonheated tipping bucket and, therefore, did not accurately record snowfall. Small snowfall amounts were recorded upon melting, but they were diminished slightly by any sublimation in the interim. Large snowfall amounts could overtop the gage and be substantially undermeasured. Because the rainfall data were used primarily to determine times when the EC sensors were affected by water, these limitations in the snowfall record were relatively unimportant. The Vaisalla HMP45C temperature-humidity probe and the RM Young 03001 wind speed and direction sensors were deployed to measure wind, temperature, and humidity data even though the EC sensors (CSAT3 and KH20) nominally measure these variables because: (1) gaps exist in the EC data record caused by precipitation or other interference; and (2) the KH20 hygrometer accurately senses rapid deviations from mean ρ_v only—it does not accurately sense mean ρ_v. The standard, non-EC sensors provided a continuous and accurate record used in troubleshooting and computing small corrections to EC data.

Table 1.   Meteorological and energy-balance sensors, locations, and data acquisition characteristics at eddy-covariance sites.

[**Abbreviations**: Bul, bulrush site; m, meters; Mix, mixed site; m^3, cubic meters; m^{-3}, per cubic meter; mm, millimeters; m s^{-1}; meters per second; W m^{-2}, watts per square meter; °, degrees; C, degrees Celsius]

Variable measured and units	Sensor make and model	Sensor height above land surface (m)	Scan interval (s)	Summary statistic(s)
Net radiation (R_n), W m^{-2}	Kipp and Zonen CNR1	Bul: 3.05; Mix: 3.28	1	Mean
Soil heat flux (G), W m^{-2}	Radiation and Energy Balance Systems HFT3	Bul: -0.02; Mix: -0.02	1	Mean
Wind speed (U_{cup}), m s^{-1}	RM Young 03011	Bul: 3.81; Mix: 3.91	1	Mean
Wind direction (WD), °	RM Young 03011	Bul: 3.81; Mix: 3.91	1	Mean, standard deviation
Soil water content (θ), m^3 water m^{-3} soil	Campbell Scientific CS616	Bul: 0 to -0.3 integrated Mix: 0 to -0.3 integrated	1,800	Sample
Rainfall (R), (mm)	Texas Electronics 525MM	Bul: 3.22; Mix: 3.56	1	Total
Air temperature (T_a), °C	Vaisalla HMP45C	Bul: 3.73; Mix: 3.91	0.1	Mean
Relative humidity (RH), percent	Vaisalla HMP45C	Bul: 3.73; Mix: 3.91	0.1	Mean
Water temperature (T_w), °C	Onset Tidbit v2	Bul: 0.01, 0.27, 0.52, water surface Mix: 0.01, 0.27, 0.52, water surface	4	Mean

Meteorological and Energy-Balance Data Processing

Saturated vapor pressure (e_s, kPa) was computed from measured air temperature (T_a) using the algorithm of Lowe (1977). This is the maximum amount of water vapor the air can hold, and it increases with increasing temperature. Actual vapor pressure (e, kPa) was computed from e_s and measured relative humidity (RH, percent) as: $e = 0.01 \times e_s \times RH$. Vapor density ($\rho_v$, g/m^3) is the mass of water vapor per unit volume of air and was computed from e using the ideal gas law: $\rho_v = M_w \times e / (R \times T_k)$ where M_w is the molar weight of water (18.0153 g/mol), R is the universal gas constant (0.0083145 kPa (m^3/mol)/K), and T_k is air temperature in kelvin.

Mean atmospheric pressure (P, kPa) was calculated using the atmospheric scale height equation: $P = P_0 \times \exp(-z / 8434.4)$ where P_0 is mean atmospheric pressure at sea level (101.33 kPa), z is site altitude (m), and 8434.4 is the scale height of the atmosphere (m). Air density (ρ_a, g/m^3) was calculated as the sum of dry air density (ρ_d, g/m^3) and vapor density (ρ_v): $\rho_a = \rho_d + \rho_v$ where $\rho_d = M_d \times (P - e) / (R \times T_k)$ and M_d is the molar weight of dry air (28.97 g/mol). Specific humidity (q, g H$_2$O/g air) is the mass of water vapor per unit mass of air, and was calculated as $q = \rho_v / \rho_a$. Specific heat of air at constant pressure, c_p (in joules per kilogram per degree Celsius), was calculated as the weighted average of the specific heats of dry air (c_{pd}) and water vapor (c_{pv}) at constant pressure: $c_p = (1 - q) \times c_{pd} + q \times c_{pv}$ (Brutsaert, 1982, p. 43). Quadratic expressions were fit to tabulated data given for c_{pd} and c_{pv} (Garratt, 1992, p. 285; Engineering Toolbox, 2011), which vary slightly as a function of temperature.

Wind speed measured with a cup anemometer and wind direction measured with a vane at 1 Hz were processed using data-logger software to compute mean wind speed (U_{cup}) and mean wind direction (WD) each half hour. Each 1-second U_{cup} and WD were combined to produce a wind vector, and the 1-second wind vectors were added in vector fashion to compute the resulting half-hour summaries.

Soil-heat flux, G, was measured directly using a soil-heat flux plate (HFT3, table 1) buried at about 2 cm beneath the soil surface. Any change in energy storage in the soil above the plate was assumed to be negligible. The plates were not buried at the beginning of the study due to the depth of standing water at that time. Scheduled site visits did not correspond to the cessation of standing water, so the plates were buried during the next visit after cessation. This late installation created a period of about 25 days in mid-2008 when G was unmeasured. Later data indicated that when water depth exceeded about 20 cm, $G \approx 0$. A model of G was created to fill in the missing data during 2008 when water depth was less than 20 cm. G was modeled as a linear combination of current and past R_n and T_a, using data from times of similar water depth during 2009 when the plate was operational. Candidate past values of R_n and T_a included 30-minute data from the previous 7 half-hours, and the mean from the past 24 hours. Multiple linear stepwise regression was used to select the subset of variables at each site that were significantly related to G. Root-mean-squared errors ($RMSE$) of the two models were 8.7 and 2.1 W/m² at the bulrush and mixed sites, respectively. Coefficients of determination (r^2) were 0.78 and 0.94, respectively.

When standing water was present, change in energy stored in the water column, Q_x, was calculated from temperatures measured at various depths in the water column. Temperature sensors were located at nominal heights of 0.01, 0.27, and 0.52 m above the soil surface, and (floating) at the water surface (table 1). These sensors divided the water column into layers and the change in heat content of each layer was computed. The value of Q_x was computed as the sum of the change in heat stored in each of the layers:

$$Q_x = \frac{C}{TP}\left(TH_1^t \times T_1^t - TH_1^{t-1} \times T_1^{t-1} + \sum_{i=2}^{n} TH_i \times \Delta T_i \right) \quad (10)$$

where

Q_x is in watts per square meter;

C is the volumetric specific heat of water (4.187 $\times 10^6$ joules per cubic meter per degree Celsius (J/m³/°C));

TH is the thickness of each layer, in meters;

T is the mean temperature of each layer, in degrees Celsius;

ΔT is the change in temperature from the previous time period for layers 2 through n, in degrees Celsius;

TP is the length of time period between temperature measurements, in seconds;

t refers to the present time period;

$t-1$ refers to the previous time period;

1   refers to the first layer;

i refers to the ith layer; and

n is the number of layers, where layers are numbered from the top down.

Each layer is bounded by temperature sensors at the top and bottom of the layer, and the mean temperature of the layer is set equal to the mean of the upper and lower sensors. This formula accounts for a change in thickness of the top layer caused by a change in lake level during the time period and assumes lower layers are constant thickness. As water level varied and fixed sensors became exposed or submerged, the value of n changed accordingly. This formula was applied to time periods of 30 min.

Gaps in the Q_x record occurred for two reasons: (1) the sensors were not deployed early enough each year to record the onset of inundation by the rising water; and (2) at times the floating sensor would become entangled in vegetation as the water level receded, causing the sensor to be suspended in air, invalidating the measurement. Models of each of the four water temperatures were devised for gap-filling, using data from periods when measurements were valid. These models are linear combinations of lagged air temperature, sensor depth, and vegetation height above water surface. Air temperature was lagged using a recursive filter, whose coefficients also depended on sensor depth. Because of small variations from year to year in actual sensor height above land surface and vegetation shading patterns, a separate model was calibrated for each sensor-year combination. The $RMSE$ of the temperature models ranges from 0.47 to 1.47°C, averaging 1.00°C, and the adjusted r^2 ranges from 0.872 to 0.984, averaging 0.938. Due to the extreme sensitivity of Q_x to small changes in temperature, coupled with the random intermittent heating of the sensors from the solar beam penetrating gaps in the vegetation canopy, both computed and modeled Q_x were quite noisy on a 30-minute time step, but they displayed reasonable daily patterns. Therefore a seven-point running mean was used as the final best estimate of both measured and modeled Q_x. The $RMSE$ of the Q_x models varies from 45.1 to 100.1 W/m², averaging 74.6 W/m², and the r^2 ranges from 0.580 to 0.826, averaging 0.739. Although these $RMSE$ values appear to be large at first glance, Q_x typically ranged from about -300 W/m² at night to about 400 W/m² in late morning in summer—a range of about 700 W/m², or about 10 times the mean $RMSE$. The main purpose of measuring and modeling Q_x on a 30-minute basis (since long-term Q_x approaches zero by definition) is for estimating missing or bad LE data using the energy-balance equation (eq. 9). The Q_x models used here are considered to be adequate for this purpose.

Biophysical Measurements and Observations

During site visits, various measurements and observations were made to characterize slowly changing variables. Mean canopy height, mean height of the dead vegetation stalk layer, and mean water depth were measured at 6 to 10 locations at each site. Vegetation height between site visits was estimated using linear interpolation. Observations of greenness (whether the canopy was live or dormant) were made to determine the potential for transpiration to occur. Many photographs were taken during each visit to substantiate the observations and measurements.

Lake stage is measured daily by the USGS at the Rocky Point station (11505800; fig. 1) and was used to determine daily water depth at each site. The land surface altitude at each site was calculated by subtracting the observed mean water depth from the lake stage during each of the nine site visits when standing water was present. Each depth observation was the mean of six to eight measurements near the platform. Site altitudes were determined to be 1,261.88 m (4,140.02 ft) and 1,262.03 m (4,140.52 ft), with standard deviations of 3.3 cm and 4.6 cm ($n = 9$) at the bulrush and mixed sites, respectively.

Bowen-Ratio Energy Balance (BREB) Measurements at Open-Water Sites

Evaporation is one of the largest components of the budget of Upper Klamath Lake (Hubbard, 1970). Past efforts to determine the evaporative flux have largely relied on evaporation pan data (Hubbard, 1970; Kann and Walker, 1999) or modeling approaches based on meteorological inputs (for example, Hostetler, 2009). Prior to the current study, the only effort to measure evaporation from Upper Klamath Lake is the work of Janssen (2005), who used an energy-budget approach similar to that employed in this study. The energy budget method provides an indirect measurement of evaporation as represented by the latent-heat flux. The energy budget method is based on the conservation of energy and the ability to measure or calculate all major energy inflows to and outflows from the lake except for the latent- and sensible-heat fluxes. By then measuring the ratio of sensible- to latent-heat flux (the Bowen ratio), the latent-heat flux and evaporation rate can be computed.

Bowen-Ratio Energy-Balance Method

The Bowen-ratio energy balance (BREB) method described by Anderson (1954) was used to calculate evaporation from Upper Klamath Lake for biweekly budget periods May through October, 2008 through 2010 (table 2). Full-year measurements were not possible because the floating instrumentation platforms were removed from the lake during winter months. The instrumentation platforms in the lake were deployed for water quality and hydrodynamic studies (Wood and others, 2006; Wood and Gartner, 2010). The two floating platforms are referred to as the midlake (MDL) site near the middle of the lake and the midlake north (MDN) site

near the middle of the northern section of the lake (fig. 1). Meteorological instrumentation at the EC stations, described in Meteorological and Energy-Balance Sensors and Data Collection, was used for supplementary data collection and gap-filling. Data used for this study collected at the MDL and MDN sites include water temperature at the surface, 1 m below the surface, and 1 m above the lake bottom; and air temperature and relative humidity at approximately 2 m above the water surface. Because no relative humidity data were available at the MDN site during 2008, evaporation rates were only calculated at the MDL site that year. However, evaporation rates at the MDL and MDN sites were both calculated in 2009 and 2010.

The energy-budget evaporation rate is determined using the following equation (from Winter and others, 2003) which can be derived from equation 3 and describes the evaporation rate as a function of components of the lake energy budget that can be measured or calculated:

$$E_{eb} = \frac{R_s - R_{sr} - R_a - R_{ar} - R_{bs} + Q_v - Q_b - Q_x}{\rho_w \left[\lambda (1+\beta) + c_w T_s \right]} \qquad (11)$$

where
- E_{eb} is the energy-budget evaporation rate (m/s),
- R_s is incoming solar radiation (W/m^2),
- R_{sr} is reflected solar radiation (W/m^2),
- R_a is incoming long-wave atmospheric radiation (W/m^2),
- R_{ar} is long-wave atmospheric radiation reflected from the water body (W/m^2),
- R_{bs} is long-wave radiation emitted from the water body (W/m^2),
- Q_v is net advected energy (W/m^2),
- Q_b is energy transferred to lakebed (W/m^2),
- Q_x is the net change in energy stored in water body (W/m^2),
- ρ_w is the density of water (1,000 kg/m^3),
- λ_v is the latent heat of vaporization of water (J/kg),
- β is the Bowen ratio (dimensionless),
- c_w is the specific heat of water (4,184 J/kg/°C), and
- T_s is the temperature of the water surface (°C)

Discussions of equation 11 can be found in Anderson (1954), Sturrock and other (1992), Winter and others (2003), and Janssen (2005). Outgoing long-wave radiation is the sum of R_{ar} and R_{bs}. The first five terms in the numerator of equation 11 constitute net radiation, R_n. Other than E_{eb}, all terms of equation 11 are known, or can be measured or calculated. Measurements and calculations involving radiation terms, advected energy, energy transferred to lakebed, and Bowen ratios are described in the following sections.

Table 2.   Biweekly budget periods used for calculations.

[For any given budget period, the first date and time (24-hour clock) is the beginning of the period and the second is the end. Calculations were not made for shaded periods due to incomplete data]

Date	Budget period	Date	Budget period	Date	Budget period
05-01-08 0:00 05-15-08 0:00	1	05-01-09 0:00	14	05-01-10 0:00 05-15-10 0:00	27
05-15-08 0:00 05-29-08 0:00	2	05-15-09 0:00 05-29-09 0:00	15	05-15-10 0:00 05-29-10 0:00	28
05-29-08 0:00 06-12-08 0:00	3	05-29-09 0:00 06-12-09 0:00	16	05-29-10 0:00 06-12-10 0:00	29
06-12-08 0:00 06-26-08 0:00	4	06-12-09 0:00 06-26-09 0:00	17	06-12-10 0:00 06-26-10 0:00	30
06-26-08 0:00 07-10-08 0:00	5	06-26-09 0:00 07-10-09 0:00	18	06-26-10 0:00 07-10-10 0:00	31
07-10-08 0:00 07-24-08 0:00	6	07-10-09 0:00 07-24-09 0:00	19	07-10-10 0:00 07-24-10 0:00	32
07-24-08 0:00 08-07-08 0:00	7	07-24-09 0:00 08-07-09 0:00	20	07-24-10 0:00 08-07-10 0:00	33
08-07-08 0:00 08-21-08 0:00	8	08-07-09 0:00 08-21-09 0:00	21	08-07-10 0:00 08-21-10 0:00	34
08-21-08 0:00 09-04-08 0:00	9	08-21-09 0:00 09-04-09 0:00	22	08-21-10 0:00 09-04-10 0:00	35
09-04-08 0:00 09-18-08 0:00	10	09-04-09 0:00 09-18-09 0:00	23	09-04-10 0:00 09-18-10 0:00	36
09-18-08 0:00 10-02-08 0:00	11	09-18-09 0:00 10-02-09 0:00	24	09-18-10 0:00 10-02-10 0:00	37
10-02-08 0:00 10-16-08 0:00	12	10-02-09 0:00 10-16-09 0:00	25	10-02-10 0:00 10-16-10 0:00	38
10-16-08 0:00 05-15-09 0:00	13	10-16-09 0:00	26	10-16-10 0:00	39

Radiation Terms

Energy budget studies commonly employ upward facing radiometers to measure incoming long- and short-wave radiation, and calculate reflected and emitted radiation (Winter and others, 2003). This study employed a net radiometer deployed over water to measure both incoming and reflected short- and long-wave radiation. For stability, the net radiometer was anchored to the lake bottom using a tripod near the lake margin. The site was selected with southern exposure to the lake to ensure reflected radiation was from the water surface only. Conditions at the site are considered reasonably representative of the whole lake. The instrument was operational July through October in 2008 and 2009 (instruments were removed from the lake during winter due to ice), hence there are substantial gaps in the measurement record.

During periods when data from the over-water net radiometer were not available, incoming radiation data from the two radiometers deployed on the eddy-covariance platforms were used. Although incoming radiation is comparable at all the sites, the reflected radiation is considerably different because the eddy-covariance sites were over a vegetated landscape. Because of this, calculated values of reflected solar radiation were used during periods of missing record. Reflected shortwave solar radiation R_{sr} is calculated as:

$$R_{sr} = R_s a \qquad (12)$$

where a is the albedo (reflectivity) of water dependent upon the solar zenith angle (φ) at the approximate center of the lake (Janssen, 2005; Oke, 1987; Lee, 1980). Solar zenith angles are estimated by equations which were adapted from a publicly accessible National Oceanic and Atmospheric Administration spreadsheet. The equation for the albedo of a smooth water surface as presented in Janssen (2005) is:

$$a = a_0 + ce^{b\varphi} \qquad (13)$$

where a_0, c, and b are constants. These constants were determined by nonlinear regression using estimated albedos, calculated as $\dfrac{Q_{sr}}{Q_s}$, from the 2008 and 2009 incoming and reflected solar radiation data at the over-water net radiation site as the regressor and corresponding solar zenith angles at the approximate center of the lake, as the regressand. Substituting albedos calculated from equation 13 into equation 12 results in half-hour reflected shortwave radiation values for open water.

Reflected long-wave radiation, R_{ar}, is calculated as 3 percent of the measured incoming long-wave radiation (Anderson, 1954; Sturrock and others, 1992; Janssen, 2005). Long-wave radiation emitted from the water body, R_{bs}, is calculated using the Stefan-Boltzmann Law for gray-body radiation (Jannsen, 2005; Dingman, 2002):

$$R_{bs} = \varepsilon\sigma T_s^4 \qquad (14)$$

where

  T_s is the temperature of the water surface, averaged from MDL and MDN sites when possible, in kelvin,

  σ is the Stefan-Boltzman constant $[5.671\times10^{-8}\ \mathrm{Wm}^{-2}\mathrm{K}^{-4}]$, and

  ε is the emissivity of the water surface [dimensionless] taken to be 0.97 (Anderson, 1954).

Net Advected Energy

Net advected energy, Q_v, for a budget period is calculated as the total energy exiting the lake through streamflow out of the lake subtracted from the total energy entering the lake through streamflow into the lake, precipitation, and groundwater flow into the lake (Anderson, 1954):

$$Q_v = \frac{c_w\rho_w}{A_L}\sum(q_{in}T_{q\ in} - q_{out}T_{q\ out}) \qquad (15)$$

where

  A_L is the area of the lake surface, in square meters,

  q_{in} and q_{out} are the flows entering and exiting the lake, in cubic meters per second, and

  $T_{q\ in}$ and $T_{q\ out}$ are water temperatures of the incoming and outgoing water flows, in degrees Celsius.

The area of the lake is determined in the same manner as Janssen (2005) using the Bureau of Reclamation 1974 Upper Klamath Lake stage-area curve (fig. 6). Areas were determined from this curve using stage data from U.S. Geological Survey lake stage gage 11507001. Stage at this station is calculated as a daily weighted mean of lake elevations from three gages in order to offset seiche effects (Janssen, 2005). Daily mean lake stage data for Upper Klamath Lake are reported to the nearest 0.01 ft while the stage-area curve is incremented by 0.1 ft. Hence, interpolation was used to determine areas for stages falling between points on the curve, to reduce rounding errors.

Sources of uncertainty in calculating advected energy entering the lake are the lack of flow and temperature data for many of the smaller streams feeding the lake. The rate and temperature of groundwater discharge to the lake is also poorly quantified. During this study, the only measured inflow to the lake was that of the Williamson River, which accounts for about half the total inflow.

Temperature data were collected for the Williamson and Wood Rivers, and Sevenmile Canal near their outlets into the lake. During low flow conditions, temperature probes periodically were exposed to air resulting in several gaps in the records. Regression techniques, using these same streams as regressors (independent or explanatory variables) and regressands (dependent or response variables) in turn, were used to gap-fill missing portions of records (table 3). This was possible since the majority of time the probes of the three streams were not out of the water at the same time. Linear interpolation was used to gap-fill the few parts of records in which regressions could not be used.

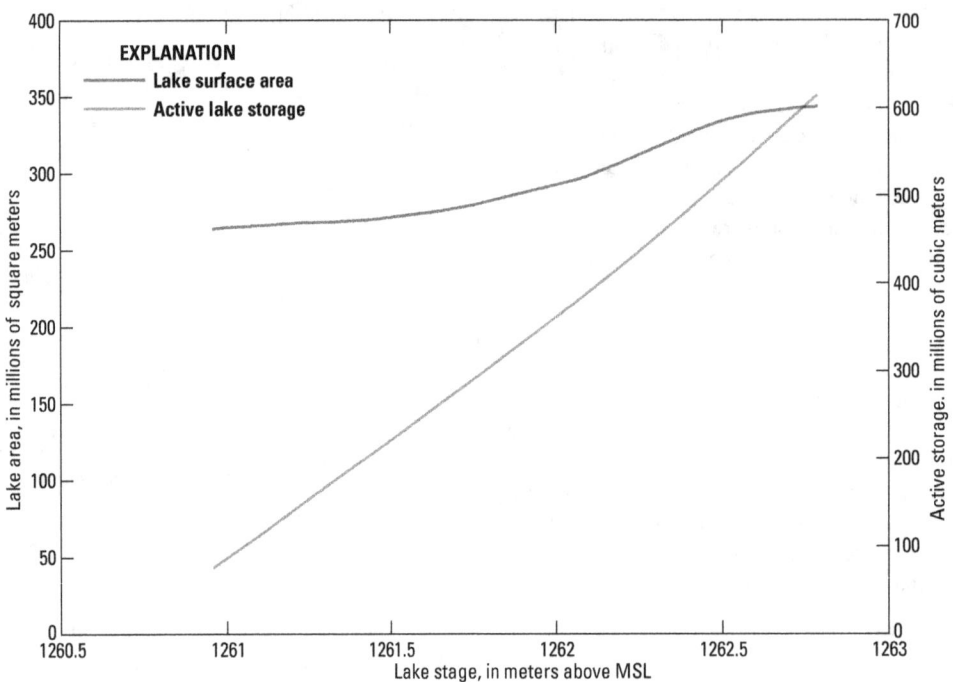

Figure 6.   Lake stage area and live storage volume curves for Upper Klamath Lake. Active storage refers to the portion of the lake above the outlet control. There is also a dead storage volume below this elevation. Total storage in the lake is the sum of active and dead storage. MSL, mean sea level above Bureau of Reclamation datum.

Table 3.   Regressions used to gap-fill Williamson River, Wood River, and Seven Mile Creek temperature records and their associated goodness of fit statistics.

[Regressions only involve these three streams and 'vs all' indicates the use of two regressors. Original temperature measurements were in degrees Fahrenheit. **Abbreviations**: n, number of observations; $\hat{\sigma}$, standard error of regression, r^2, coefficient of determination]

Regression	n	$\hat{\sigma}$	r^2
Williamson vs all	290	1.5	0.93
Wood vs all	290	1.3	0.92
Seven vs all	290	3	0.85
Seven vs Wood	379	3.4	0.82
Williamson vs Wood	311	1.7	0.93

To account for the ungaged flow into the lake, an approach was developed which centered on a simple mathematical model of the lake's water budget. For this approach, we subtract the gaged inflow from estimates of the total outflow considering estimated evaporation and measured changes in lake volume to estimate the ungaged inflow for budget periods as shown in equation 16:

$$Q_{out} + E + \Delta V - Q_{in,g} = Q_{in,un} \qquad (16)$$

where

Q_{out} is the average outflow exiting the lake over a budget period,

E is estimated average evaporation lost to the atmosphere from the lake surface,

$Q_{in,g}$ is the average gaged inflow entering the lake, and

$Q_{in,un}$ is the resulting estimation of average ungaged inflow over a budget period.

For this equation, the Jensen-Haise method described in Rosenberry and others (2007, table 1) is used to estimate evaporative rates, and these rates multiplied by lake surface area from the stage-area curve produce volumetric estimates of evaporation for a budget period. Changes in lake volume for a budget period are derived directly from the Bureau of Reclamation stage-volume curve.

With the ungaged inflow now estimated for budget periods, equation 15 requires a means to partition these flows to the various ungaged components that contribute to the lake. This is accomplished using the detailed water budget of Hubbard (1970) which includes measurements of inflow from all significant sources during a 3-year period from 1995 through 1997. Using Hubbard's monthly values, mean monthly coefficients were calculated for the discharge of ungaged components to the lake. Coefficients for budget periods were then determined using weighted averages of mean monthly coefficients and the number of days of a particular month in a time period. Finally, the ungaged inflow of a budget period was multiplied by a stream's budget period coefficient to yield that portion of the flow contributed to the lake by the stream. Sevenmile Creek temperatures were assigned to all ungaged components except groundwater, which was estimated at a constant temperature of 10°C for all budget periods.

Energy entering the lake by precipitation was based on Bureau of Reclamation precipitation gages and air temperature sensors at the AGKO and KFLO AgriMet sites, just to the north and south of the lake respectively. As with a previous study of the lake, the temperatures of precipitation at the two stations were assumed equal to the air temperatures at the AgriMet sites (Janssen, 2005). Temperatures and precipitation from the two stations were averaged in an attempt to represent mean conditions across the lake.

Advected energy exiting the lake is calculated based on the discharge of Link River, diversions through the A-Canal, and the temperature recorded at the Link River gage. The temperature of water diverted into the A-Canal is assumed equal to that of Link River.

Energy Transferred to Lakebed

Daily Q_b is calculated by parameterization of heat flow and temperatures of water overlying the lakebed (Janssen, 2005; Pearce and Gold, 1959):

$$Q_b = \sqrt{\frac{2\pi}{p\alpha}} KA \sin\left(\frac{2\pi t}{p} + \frac{\pi}{4} + c\right) \qquad (17)$$

where

p is the period of temperature variation [365 days],

α is the thermal diffusivity of the lakebed sediments, which is assumed equal to the value of gyttja soils presented by Wetzel and Likens (1991) [$1.728 \times 10^{-2}\,\text{m}_2\,\text{d}^{-1}$,

K is the thermal conductivity of the lakebed sediment [$0.75\,\text{Wm}^{-1}\,^{\circ}\text{C}^{-1}$] as estimated by Sass and Sammel (1976),

A is the amplitude of the water temperature variation overlying the lakebed, in degrees Celsius,

t is the time, in days, beginning June 1st for a year of interest, and

c is the phase offset, in radians, of the observed temperature wave.

Amplitude (A) and phase (c) are estimated by means of a four-parameter sine wave of the form:

$$Tb(t) = Tb_0 + A \sin\left(\frac{2\pi t}{365} + c\right) \qquad (18)$$

where

$Tb(t)$ is the lake bottom temperature at time t, and

Tb_0 is the mean-annual lake bottom temperature.

This is fitted by means of sinusoidal regression to the lake bottom temperature (fig. 7), based on the average of the temperatures measured 1 m above the lake bottom at water-quality stations MDN, EPT, SET, and MDT (fig. 1; T.M. Wood, U.S. Geological Survey, unpub. data). Janssen (2005) found no statistical difference between daily lake bottom temperatures and corresponding Link River temperatures. Therefore, if no records were available from any of the water-quality stations for a day or period of days, daily mean temperatures at the Link River were used.

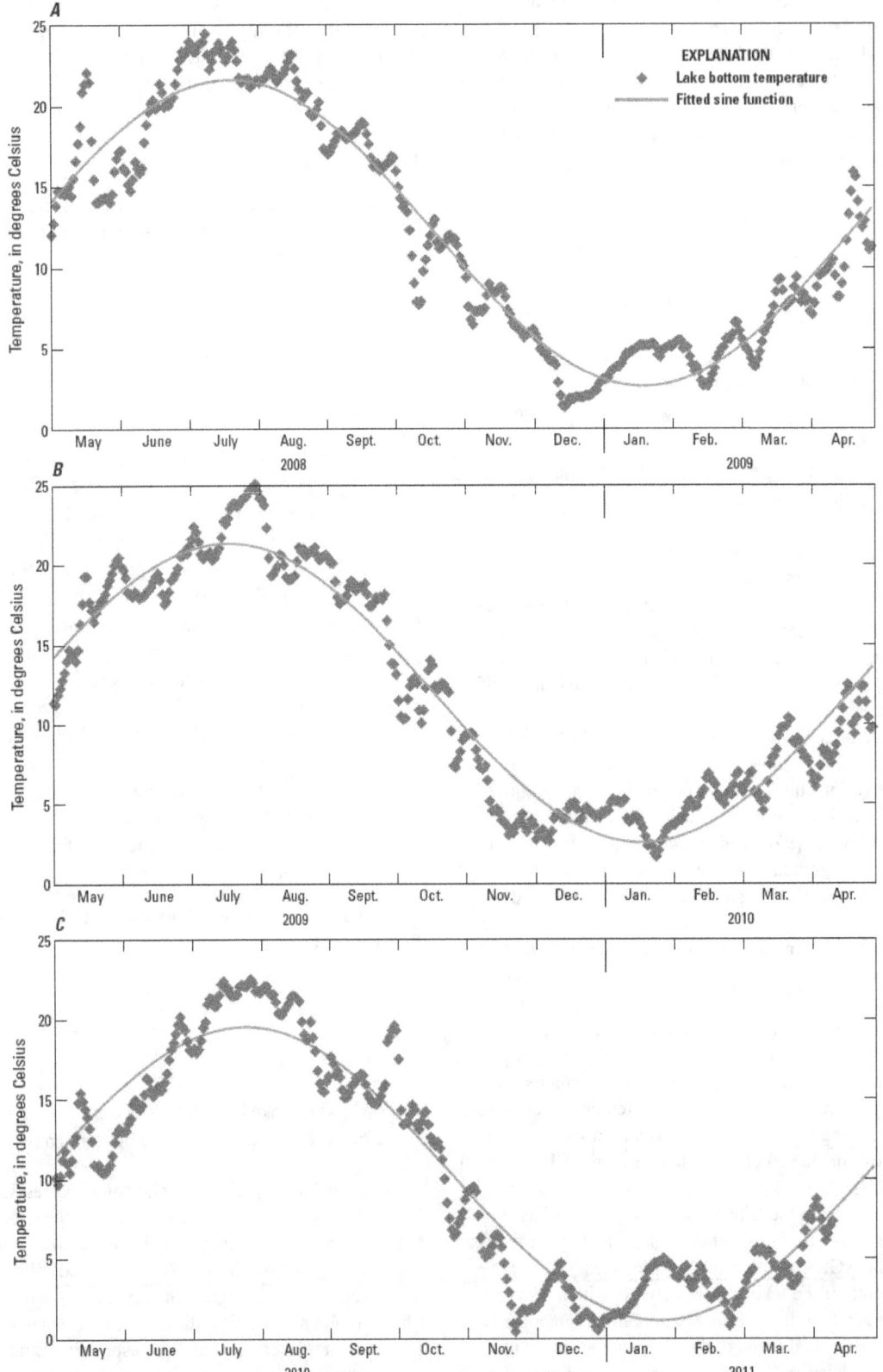

Figure 7. Upper Klamath Lake bottom temperatures and fitted sine function used to parameterize amplitude (A) and phase (c) during *A*, 2008, *B*, 2009, and *C*, 2010.

Changes in Energy Stored in the Lake

Changes in energy stored in the Upper Klamath Lake were estimated using the same well-mixed lake assumption that Janssen (2005) used for his study. This assumption is based on the findings of Wood and others (2006), who demonstrated that Upper Klamath Lake does not develop a strong thermocline due to its shallow depth. When thermal stability does develop, it typically disappears in less than a day (Wood and others, 2006). The following equation is used to estimate the changes in stored energy:

$$Q_x = \frac{c_w \rho_w}{A_L t}\left(T_i V_i - T_f V_f\right) \tag{19}$$

where

t	is the total time of the budget period, in seconds,
T	is the average temperature of the water column, in degrees Celsius,
V	is the volume, in cubic meters, of the lake, and
i and f	indicate the initial and final value of a budget period, respectively.

Temperature data collected at the five temperature stations on the lake (fig. 1) 1 m below the water surface and 1 m above the lake bottom were used to calculate an average water column temperature at that location. Average water column temperatures for a particular day were then averaged to estimate the mean temperature of the lake. Lake stage data from USGS gage 11507001 and storage volume, interpolated to nearest 0.01 ft (fig. 1), were used for calculations in equation 19. Dead storage of 2.61×10^8 m^3 (211,300 acre-ft; from the station description for the above referenced gage) was added to the active storage term to yield total lake volume.

Bowen Ratio

The Bowen ratio (β in the denominator of equation 11) is the ratio of sensible heat to latent heat. It is calculated (Bowen, 1926; Sturrock and others, 1992) as:

$$\beta = \gamma\left(\frac{T_s - T_a}{e_s - e_a}\right) \tag{20}$$

where

T_s and T_a	are the water surface and air temperatures respectively, in degrees Celsius,
e_s and e_a	are vapor pressures at the water surface and air, respectively, in kilopascals, and
γ	is the psychometric constant, in kilopascals per degree Celsius.

The psychometric constant is a function of the specific heat of air c_a [1.00416 J kg^{-1} °C], atmospheric pressure P [kPa] and the latent heat of vaporization λ_v (Janssen, 2005; Dingman, 2002):

$$\gamma = \frac{c_a P}{0.622 \lambda_v} \tag{21}$$

The latent heat of vaporization [J kg^{-1}] is a function of surface temperature and can be approximated by (Janssen, 2005; Dingman, 2002):

$$\lambda_v = 2.49 \times 10^6 - 2.36 \times 10^3 T_s \tag{22}$$

Atmospheric pressure was estimated by (Janssen, 2005; Shuttleworth, 1993):

$$P = 101.33\left(\frac{293 - 0.0065z}{293}\right)^{5.256} \tag{23}$$

where

z	is lake stage altitude, in meters (USGS station station 11507001).

Vapor pressure of the air is estimated as:

$$e_a = RH \times 0.611 e^{\frac{17.3 T_a}{T_a + 237.3}} \tag{24}$$

where RH is the relative humidity [dimensionless] and the rest of the right-hand side is the saturated vapor pressure of the air [kPa] which is a function of air temperature (Janssen, 2005; Dingman, 2002). Similarly, the vapor pressure at the water surface can be calculated as:

$$e_s = 0.611 e^{\frac{17.3 T_s}{T_s + 237.3}} \tag{25}$$

As can be seen from equations 24 and 25, the vapor pressure at the surface is saturated and a function of water surface temperature.

One Bowen-ratio value was calculated for each 2-week energy budget period using air and water surface temperatures and vapor pressures (eq. 20) averaged over the period.

Determination of Reference Evapotranspiration and Crop Coefficients

Potential evapotranspiration (ET_p) is the ET that occurs from an extensive land or water surface under a given set of meteorological conditions when the surface is well supplied with water. In the most general interpretation, the concept applies to a wide range of land-surface types, from dense rainforests to bare soil. Because vegetation type, density, and vigor have a strong effect on the resulting ET_p, the need gradually arose for a more specific quantity—one that assumes a standard vegetation type and density, growing vigorously, and therefore depends only on weather conditions. Penman (1948) suggested dense grass cover as the reference vegetation, although other crops (and open water) have been suggested over the years. Computation of ET_p for a specific reference crop led to the adoption of the notation ET_r to refer to potential ET from a reference crop. During the late 20th century, the Food and Agricultural Organization (FAO) of the United Nations took the lead in standardizing the computation of ET_r, most recently through the work of Allen and others (1998), who presented the Penman-Monteith equation in a user-friendly form, for use with a 12-cm tall, dense grass reference crop. Using this methodology, ET for other well-watered crops is computed as the product of ET_r and a crop coefficient, K_c, which varies based on growth stage of the crop:

$$ET = K_c \times ET_r \qquad (26)$$

Subsequently, the American Society of Civil Engineers modified the FAO equation to also accommodate alfalfa as a reference crop (Allen and others, 2005). The Bureau of Reclamation maintains a network of weather stations in the Pacific Northwest region states that can be used to calculate ET and crop ET using the Allen and others (2005) alfalfa (tall crop) equation. Daily values of ET and crop ET are posted on the Bureau of Reclamation Web site (http://www.usbr.gov/pn/agrimet/h2ouse html) in near real time, for use by irrigators and others.

In the current study, we divide measured values of ET from the two wetland sites by posted values of tall crop ET_r calculated from nearby weather station data to compute daily and biweekly values of K_c. Daily values of ET_r are obtained from the Agency Lake (AGKO) weather station, located about 7.1 km northeast of the bulrush site. When daily values of ET_r are equal to 0, K_c is infinite, and we arbitrarily set K_c to a value of 6. Daily values of ET and ET_r are aggregated into biweekly values to compute biweekly values of K_c. To investigate relations between K_c and precipitation, daily values of precipitation were retrieved from the Klamath Falls (KFLO) weather station, located about 44 km southeast of the bulrush site, in Klamath Falls, Oregon. This station deploys a weighing precipitation gage, which measures snowfall more reliably than the unheated tipping bucket gage at the (closer) AGKO station. Growing-season ET data from alfalfa and pasture also were retrieved from the KFLO station for comparison with measured wetland ET. These ET values were not available from the AGKO station Web site.

Evapotranspiration Results from Bulrush and Mixed Vegetation Sites

Site Conditions Affecting Evapotranspiration

Peak vegetation heights typically were 1.9 to 2.3 m above land surface (fig. 8). Near-peak heights usually occurred from early July to early October, forming somewhat of a plateau of canopy height during the heart of the growing season. The bulrush and cattail typically senesce in late September, turning from green to brown, and then lodge over in October and November in response to wind and snow loading. The bent-over plants merge with and are supported by plants from previous years, creating a loosely woven mat of dead plants roughly 0.5 to 1 m in height, which becomes the understory of the next year's live canopy (fig. 8). In the spring, new shoots sprout from the soil, up through the water column and understory, until they emerge into the open space above. Upon emergence, the plants grow rapidly to their peak height.

Water level typically fluctuates about 1.3 m annually in response to inflows and controlled outflow at the dam, resulting in water levels that vary both above and below land surface at both sites at times of the year (fig. 8). Minimum water levels usually occur in October, and maximum water levels occur from April through June. During the study period, maximum water level at the bulrush site was 0.89 m above land surface, and minimum water level was -0.63 m. Land surface is 0.15 m higher at the mixed site than at the bulrush site, so maximum and minimum water levels at this site were 0.15 m lower (fig. 8). During 2008 to 2010, the hydroperiods ranged from 7.2 to 5.2 months and from 5.9 to 3.5 months at the bulrush and mixed sites, respectively. Minimum hydroperiods occurred in 2010 due to unusually low water levels, and hydroperiods were substantially shorter than in 2008 and 2009.

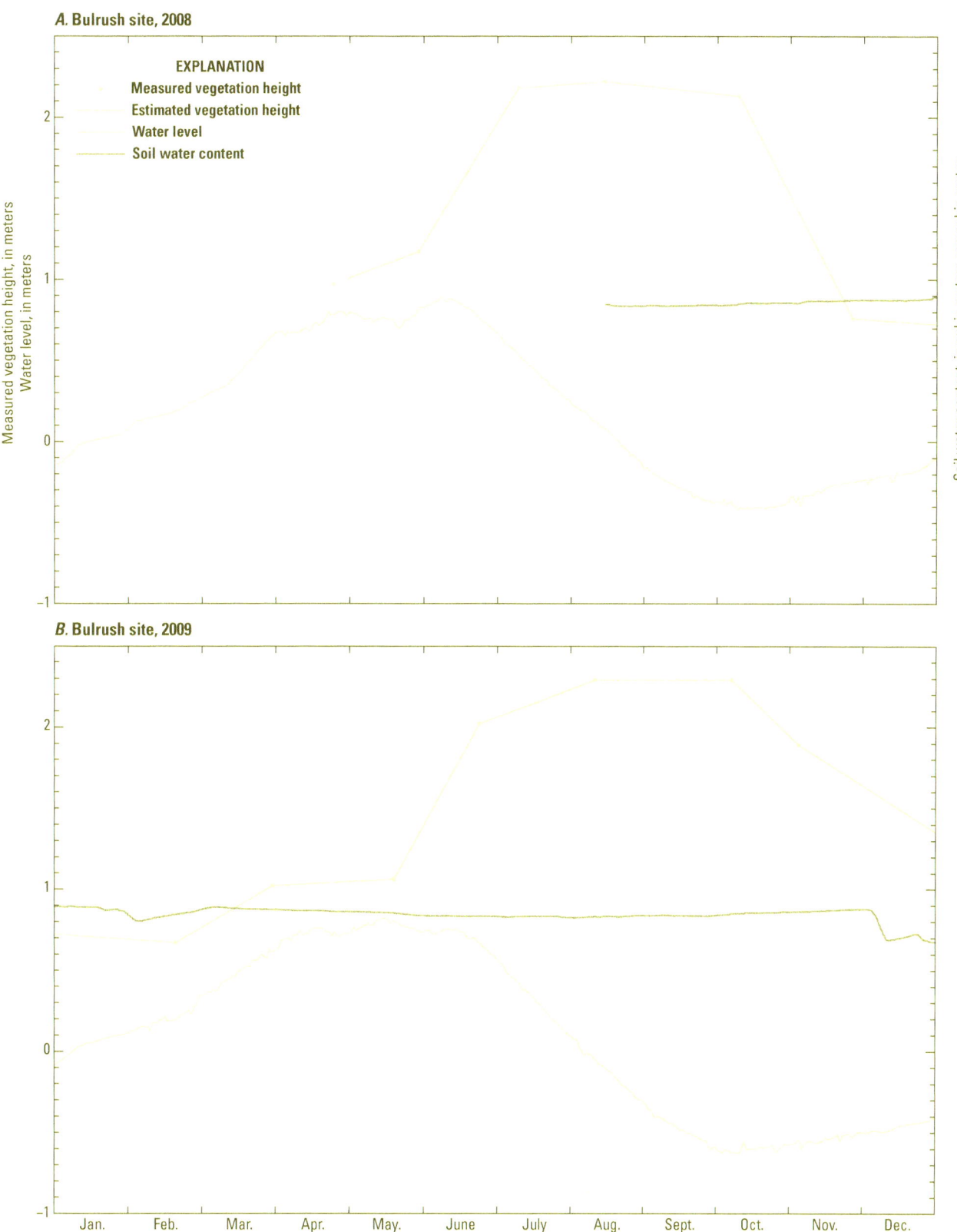

Figure 8.   Vegetation height, water level, and soil water content at bulrush site during *A,* 2008, *B,* 2009, and *C,* 2010, and at mixed site during *D,* 2008, *E,* 2009, and *F,* 2010.

C. Bulrush site, 2010

D. Mixed site, 2008

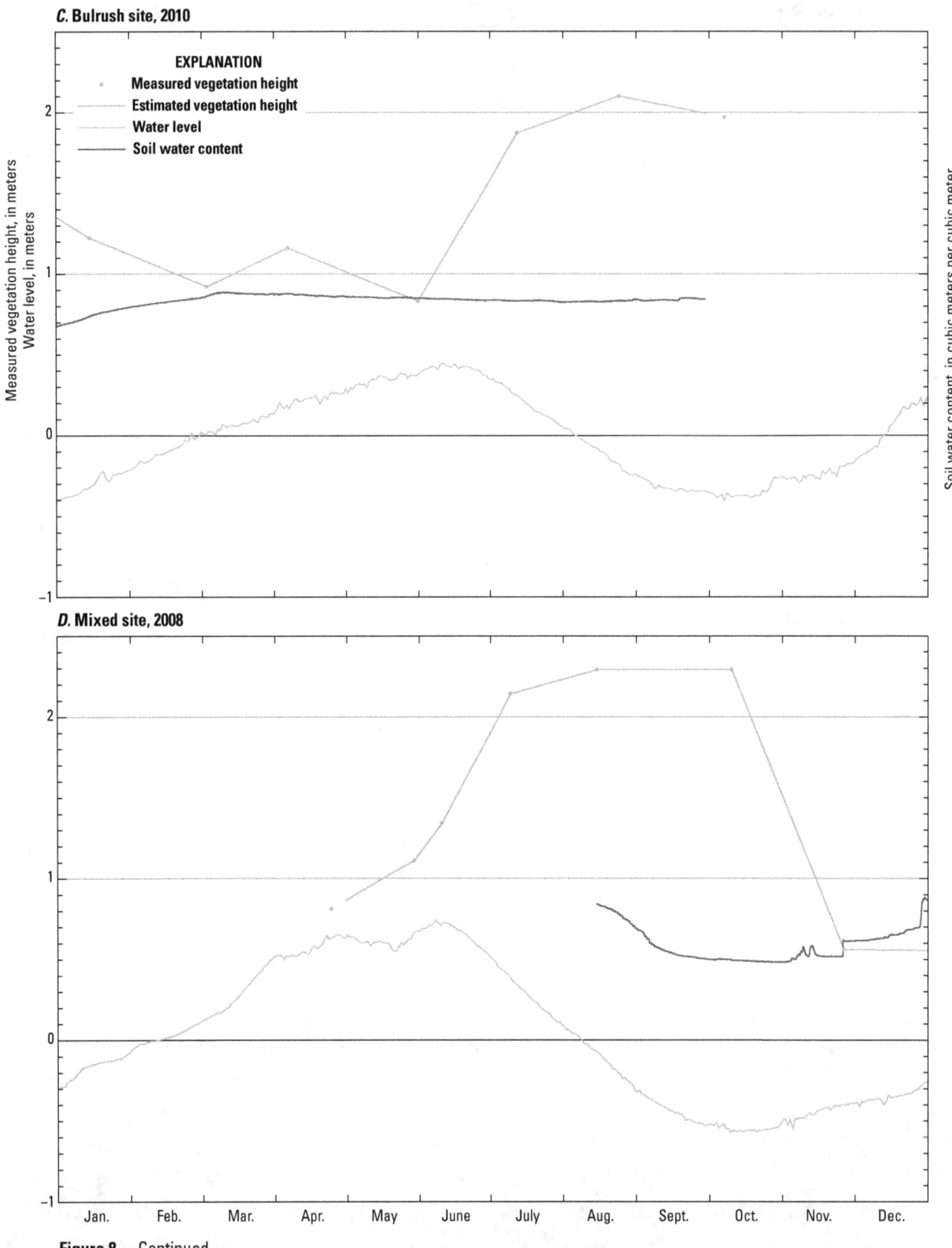

Figure 8.—Continued

E. Mixed site, 2009

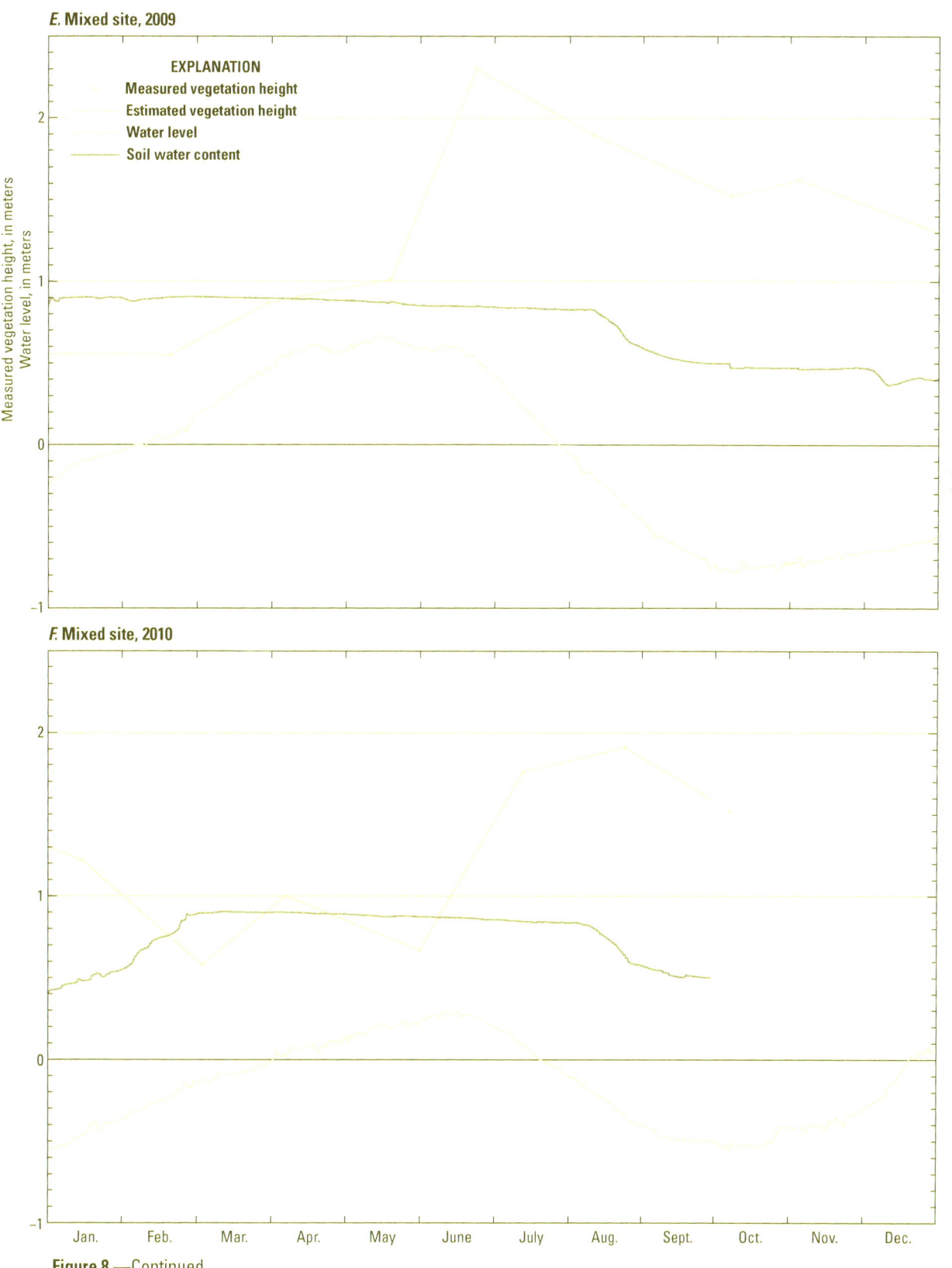

Figure 8.—Continued

Soil water content (θ; fig. 8) was measured using a 30-cm long time-domain reflectometer (TDR) installed vertically in the soil, which sensed the mean θ from the surface to a depth of 30 cm. The TDR probe was not specifically calibrated for this soil, so the absolute readings are approximate although differences should be qualitatively correct. At both sites, the probe was installed in mid-August 2008, about when the water level had receded to the land surface. During most of the study period, θ at the bulrush site was very steady at about 0.82 to 0.9 m^3 m^{-3}, a value typical of saturated peat (Schedlbauer and others, 2011). During the coldest part of both winter periods, the apparent value dropped below this range (fig. 8), but this was very likely an artifact of freezing temperatures over at least part of the TDR measurement depth interval. The TDR responds to the dielectric constant of the surrounding (moist) medium, and the dielectric constant of ice is about 4 percent that of liquid water. The low readings probably were not caused by conditions of partial saturation because similarly low readings did not occur at the times of minimum water levels each year when partial saturation would be most expected (fig. 8). The apparently saturated soil ($\theta = 0.85$) during the time of minimum water level (-0.63 m) in October 2009 suggests that the capillary fringe of the peat soil at this site is at least 0.63 m, an unusually high value. In contrast, the TDR probe at the mixed site appeared to indicate partial dewatering during all 3 years, beginning in August, when water level dropped below about 0.2 m below land surface (figs. 8D–F). Freezing soil water could not have been a factor at this time of year, and the declines in θ mirror declines in water level, suggesting a valid instrumental response and a more normal capillary fringe height of less than about 0.2 m. Soils were not sampled at the two sites, but these substantially different estimates of capillary-fringe height suggest that the soils differ greatly in mean particle size. In addition, while the saturated soil at the bulrush site indicates that the vegetation there was never stressed, the decrease in θ at the mixed site to values around 0.5 in September (figs. 8D–F) suggests that vegetation there may have experienced mild water stress toward the end of the growing season each year.

Source Areas and Fetch Considerations of Wetland Evapotranspiration Sites

A source-area model (Schuepp and others, 1990) was used to estimate the degree to which measured fluxes represent fluxes from the wetland surfaces (surfaces of interest) and to what extent they are affected (contaminated) by fluxes from the upwind open-water or other surfaces. Sensors used in the EC method must be placed at least 1.5 times the vegetation canopy height above the land (or water) surface to avoid measurement artifacts from underlying heterogeneities. Consequently, air passing by the EC sensors comes from upwind and, therefore, contains attributes (w, T_a, ρ_v) that correspond to fluxes from upwind surfaces. The source-area model describes the relative contributions to the measured flux signal from upwind surfaces as a function of upwind distance. The union of all contributing surfaces is known as the source area. The distance to the farthest upwind extent of uniform surface of interest is called the fetch.

At both wetland sites, sensor height, canopy height, and fetch were used to determine the percentage of the measured flux signal originating from the wetland surface. In this context, sensor and canopy heights refer to height above land surface or height above water surface if standing water is present. Canopy height measured at each site was assumed to be representative of the whole source area. Source areas are smallest in unstable conditions ($H > 0$, typically occur during daytime), somewhat larger in neutral conditions ($H \approx 0$, typically occur in heavy overcast, high winds, or near sunrise and sunset), and largest in stable conditions ($H < 0$, typically occur during nighttime). Neutral conditions were assumed for these calculations to produce the most conservative (largest) daytime source areas. Nighttime (stable) conditions were not evaluated because very little ET occurs at night. Source areas also vary in size depending on the canopy roughness length (z_m) and displacement height (d). The values of z_m and d were calculated as 0.1 h and 0.65 h, respectively, where h is canopy height in meters (Campbell and Norman, 1998, p. 71). In general, source area decreases with increasing canopy height. At each site, the source-area model is used to compute the percentage of the measured flux signal originating from within the wetland in the minimum fetch direction and in the predominant wind directions for both the minimum and maximum source areas occurring during the study. From these bounding and predominant conditions, estimates are made of the mean percentage of flux signal originating from within the wetland.

Minimum fetch at the bulrush site was 2.0 km at azimuths of 360° and 180° (due north and due south, fig. 1). Maximum source area occurred during the first day of data collection (May 1, 2008) when canopy height (h) was 0.20 m above 0.80 m of standing water (fig. 8A). For these conditions, the source-area model indicated that 96.6 percent of the flux signal was contained within 2.0 km of the station. Near-maximum source areas typically occurred during the winter, when canopy height was low. The minimum source area occurred at maximum canopy height (2.3 m), typically in summer and early fall (figs. 8A–C). In these conditions, 99.0 percent of the flux signal was contained within 2.0 km of the station. Predominant daytime wind directions (both during the growing season and year-round) were 305°, 360°, and 135°, in decreasing order. Fetches in the 305° and 135° directions were 5.1 and 3.0 km (fig. 1), affording 98.6 and 97.7 percent containment of the maximum source area flux signal, and 99.6 and 99.3 percent containment of the minimum source area flux signal, respectively. Considering the range and likely prevalence of these containment values, the mean time-weighted containment of the flux signal by dense wetland bulrush is estimated to be about 98 to 99 percent.

Minimum fetches at the mixed site were about 0.97 km at 90° (or due east) and 1.03 km at 240° (fig. 1) and these azimuths also corresponded to the two equally predominant wind directions. Minimum fetch is, therefore, estimated as the average of these two, or 1.0 km. Maximum source area occurred during the first day of data collection (May 1, 2008) when canopy height (h) was 0.20 m above 0.65 m of standing water (fig. 8D). Near-maximum source areas typically occurred during the winter, when canopy height was low. For these conditions, the source-area model indicated that 92.4 percent of the flux signal was contained within 1.0 km of the station. The minimum source area occurred at maximum canopy height (2.30 m; figs. 8D–E). In these conditions, 97.8 percent of the source area was within 1.0 km of the station. During average growing season conditions, 95.6 percent of the source area was within 1.0 km of the station. In addition, wind directions that afforded 1.5 km or more of fetch occurred during about 23 percent of the growing season, which provided 97.0 percent or greater containment of the source area. Considering the range and likely prevalence of these containment values, the mean time-weighted containment of the source area by dense wetland mixed vegetation is estimated to be about 95 to 96 percent.

The above estimates of flux-signal containment by the bulrush and mixed vegetation types indicate that the measured fluxes overwhelmingly originated from the surfaces of interest—and not from other, dissimilar surfaces farther upwind. In addition, the surfaces upwind of the study sites in the minimum-fetch directions consisted of open water at the mixed site and open water to the south or intermittently flooded wetland to the north at the bulrush site. The hydroperiod of the wetland to the north was similar to that of the bulrush site. During the growing season, ET from these surfaces is roughly equal to ET from the surfaces of interest, and therefore would have an insignificant effect on measured ET. The greatest potential for contrast in ET occurs in the fall and early winter, when the wetland vegetation is dormant and standing water (from rising lake level) is not yet present at the study sites. If the adjacent lake surface is unfrozen during this time, it would slightly inflate the measured ET at the sites. However, because the measured winter-time ET is about 10 percent of its annual mean value and the lake surface was frozen during part of this time, the net effect on the annual mean is very small. In summary, contamination of the measured ET from surfaces surrounding the wetland surfaces of interest is very small and probably much smaller than the inherent accuracy of the EC method.

Energy-Balance Closure and Trends

A single value of EBR was used to correct H_m and LE_m in this report to avoid the occurrence of unrealistic EBR values that can occur over shorter time periods. Correction on a 30-minute, daily, or biweekly basis is problematic because the smaller signal-to-noise ratio can produce very large or small values of EBR at times when TF and AE are small (near sunrise, sunset, and at night for 30-minute values, during overcast winter periods for daily and biweekly values), potentially producing unreliable final values of H and LE. For example, if EBR is small and values of H_m and LE_m are of similar magnitude and of opposite sign, they will both be inflated to very large numbers; and if H_m exactly equals $-LE_m$, division by zero will occur. To compare the use of a single EBR with shorter-period EBR, a biweekly EBR value was computed, and is shown along with biweekly AE, uncorrected TF, and the mean EBR at the bulrush site in figure 9. Biweekly EBR is relatively constant during the high-flux time of year: March 1 to October 31. During this time, biweekly EBR ranges from 16 percent greater than, to 12 percent smaller than the mean EBR (equal to 0.730), and averages 0.753, or 3.2 percent greater than the mean EBR. This high-flux EBR follows a subtle but consistent pattern from year to year, resembling the letter m, with maxima around mid-May and early September, and minima around mid-March, early July, and late October.

During low-flux times of the year, biweekly EBR is more variable, unpredictable, and tends to be smaller than the mean EBR (fig. 9). Over half are smaller than 60 percent, and one value drops as low as -332 percent in December, 2008, when $AE > 0$ and $TF < 0$. In addition, inter-annual variability is much greater than during high-flux times. The unpredictable and large variation of EBR during low-flux times probably is caused by a low signal-to-noise ratio in the AE and TF measurements, because those fluxes are small; less than 70 W m^{-2}, and averaging around 32 W m^{-2}. Because the low signal-to-noise ratio creates somewhat random variations in EBR, biweekly correction of TF to satisfy these EBR values is considered questionable.

The m pattern evident during the high-flux times extends into the low-flux times, although with less definition and more variability. The reason for this pattern was investigated using graphical and regression analyses, but no significant relations to other variables were found. The value of EBR has frequently been related to friction velocity, $u^* = (\tau/\rho_a)^{0.5}$, citing reduced turbulence at low wind speeds as a cause of low EBR (for example, Wilson and others, 2002; Aubinet and others, 2012,). However, EBR is virtually unrelated to u^* in this study. Using all biweekly data, EBR is not significantly related to u^* ($r = -0.318$, a negative correlation, opposite to that observed elsewhere, and $p = 0.21$). After removing the outlier point at $EBR = -332$ percent, r equals 0.187, and p equals 0.14, still an insignificant relation. All biweekly EBR results for the mixed site (data not shown) are very similar to those for the bulrush site.

Considering the relatively small errors incurred during the high-flux times by the use of a single, mean EBR, and the difficulty and questionable validity of using a biweekly EBR during low-flux times, a single, mean value of EBR was chosen to correct the TF in this study. Although ET computed this way may be slightly overestimated or underestimated at times compared to the use of a biweekly EBR, the study-period averages are identical.

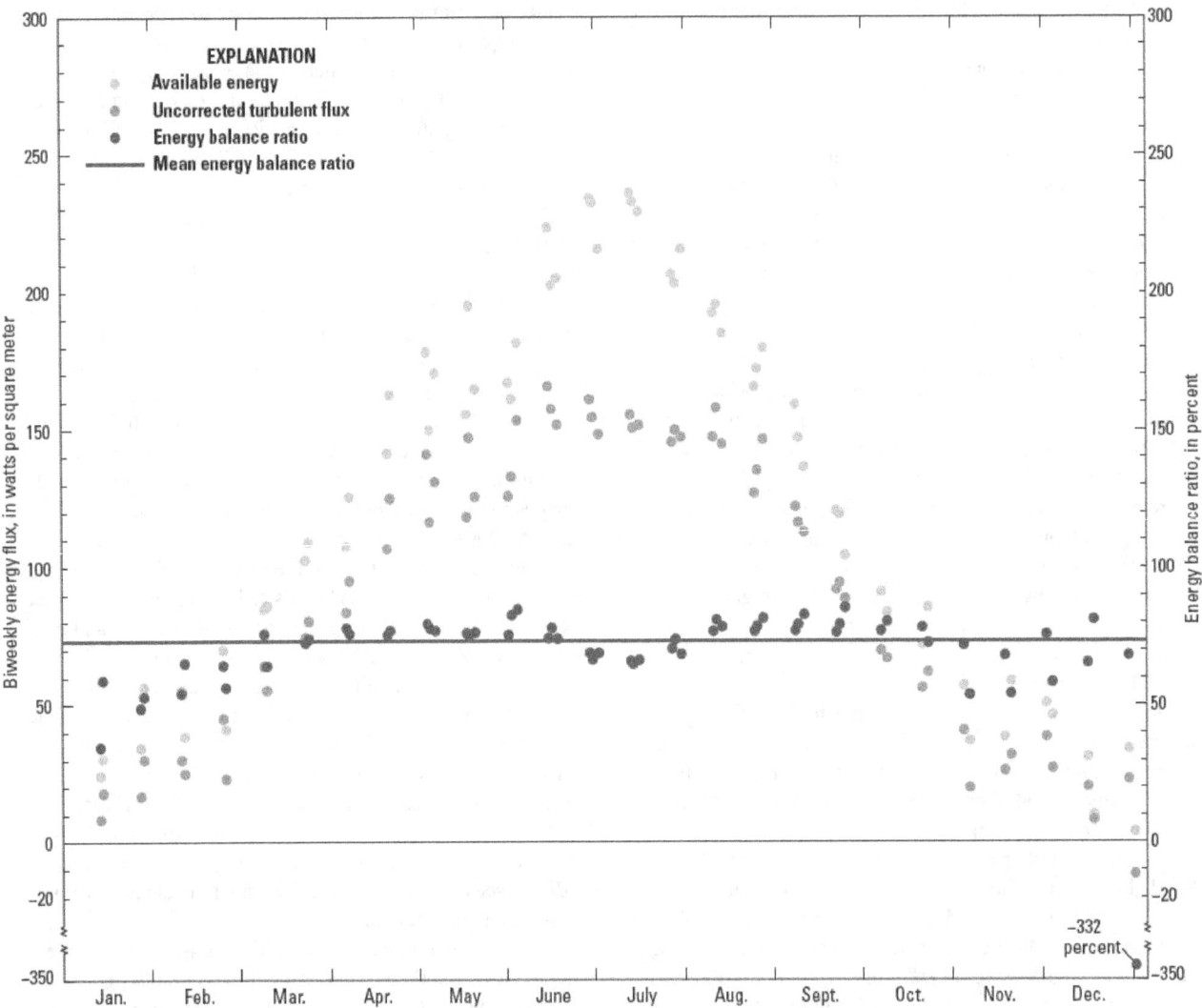

Figure 9. Biweekly means of available energy, turbulent flux, and energy-balance ratio (*EBR*) at the bulrush site during the study period. Also shown is the mean study-period *EBR*, equal to 73.0 percent.

Values of *EBR* also were computed for six different but overlapping full 2-year periods during the study to explore whether the *EBR* calculated for the whole study period (2.41 years) was representative of an annual average value. The first 2-year period began on the first day of the study, and each subsequent 2-year period began 4 weeks later the preceding one. The *EBR*s of the six 2-year periods varied by less than ± 1 percent, and the means were within 0.25 percent of the full-study period *EBR* at both sites (data not shown), indicating that the full-study period *EBR* adequately approximated the annual *EBR*, and the annual *EBR* was relatively constant during the study period.

Eddy-covariance sensors functioned correctly during 84 percent of the study period (table 4), substantially better than in higher rainfall locations such as Florida, where EC sensors were operational only 51 percent of the time during a recent study (Schedlbauer and others, 2011). The energy-balance ratio (*EBR*) was calculated from equation 8 using only data collected when all energy-balance sensors functioned correctly, constituting 63.0 and 70.8 percent of the study period at the bulrush and mixed sites, respectively (table 4). The *EBR*s of 0.730 and 0.781 at the bulrush and mixed sites, respectively, are slightly below the mean observed *EBR* of about 0.8 (Twine and others, 2000; Wilson and others, 2002; Foken, 2008), but are well within the range typically seen in many studies (Dugas and others, 1991; McCaughey and others, 1997; Twine and others, 2000; Mauder and others, 2006; Foken, 2008). Measured values of turbulent flux (H_m and LE_m) were divided by these ratios to obtain final values, designated as H and LE. Regressions between

Table 4. Percentage of time eddy covariance sensors operating (% $Time_{EC}$), percentage of time all energy-balance sensors operating (% $Time_{ALL}$), energy-balance ratio (*EBR*), root-mean-squared error between 30-minute available energy and adjusted turbulent flux (*RMSE*), and coefficient of determination (r^2) of 30-minute, daily, and biweekly means of available energy and turbulent flux.

[Only data collected when all energy-balance sensors functioned correctly were used to compute 30-minute r^2, whereas daily and biweekly r^2 values also include gap-filled data. **Abbreviation:** W m^{-2}, watts per square meter]

Site	% $Time_{EC}$	% $Time_{ALL}$	EBR	RMSE (W m^{-2})	30-minute r^2	Daily r^2	Biweekly r^2
Bulrush	84.2	63.0	0.730	71.9	0.878	0.928	0.973
Mixed	84.4	70.8	0.781	63.7	0.905	0.959	0.981

30-minute available energy and final turbulent flux yielded *RMSE*s of 71.9 and 63.7 W m^{-2} which, when compared to a range in available energy of about 1000 W m^{-2}, resulted in reasonably large r^2 values of 0.878 and 0.905 (table 4). Daily and biweekly r^2 values (using both measured and gap-filled data) increased substantially (table 4), illustrating that much of the error in 30-minute energy-balance data is random, and decreases over longer periods.

Daily values of the main energy-balance components (R_n, *H*, and *LE*) are shown in figure 10. *G* is not shown because the daily mean value (0.9 W m^{-2}) and standard deviation (17 W m^{-2}) are both quite small and *G* contributes little to the seasonal changes in energy-balance partitioning. A high degree of correlation can be seen between equivalent fluxes at the two sites, substantiating the field measurements, which were made independently of each other. The upper envelope of R_n corresponds to clear skies and displays the usual sinusoidal shape, with intermittent excursions downward caused by the occurrence of clouds. Occasionally, R_n exceeds the upper envelope sine curve when clouds are near, but not obscuring, the direct solar beam arriving at the site. These clouds can forward-scatter short-wave radiation (Monteith and Unsworth, 1990), raising R_n above clear-sky values.

The available energy delivered to the surface is partitioned between *H* and *LE*, and this partitioning varies considerably during the year as the wetland progresses through its annual life cycle (fig. 10). Partitioning is described in terms of the ratio of *H* to *LE*, known as the Bowen (1924) ratio, β. At the beginning of the year, the vegetation canopy is dormant and water level is below land surface. The loosely woven mat of dead stalks forms a complex surface, at times partially covered with snow and ice and partially bare, and at other times entirely bare. Snow and ice-covered surfaces contribute to *LE*, whereas bare stalks contribute to *H*. On average, *H* and *LE* are roughly equal during winter (β ≈ 1), alternating positions of dominance as precipitation coats the surfaces, then is redistributed and removed by evaporation and sublimation (fig. 10). During March, April, and early May, water level rises but is mostly shaded by the mat of dead stalks, and therefore has a minor effect on the equal partitioning. Both *H*

and *LE* steadily grow in response to increasing R_n. The new vegetation begins to emerge from the dead stalk mat in May or June (fig. 8), which begins to shift the partitioning toward *LE*, reducing β (fig. 10). By mid-June, *H* begins to decrease (even though R_n is still approaching peak values) due to increased transpiration from the vigorously growing canopy. This shift toward greater *LE* was somewhat delayed in 2010, a year of unusually low water levels early in the growing season. Because ample root-zone water was available for transpiration during this time, a possible alternate mechanism for the delay in 2010 is discussed in Daily Evapotranspiration and Crop Coefficients.

By midsummer, energy is partitioned overwhelmingly to *LE* (fig. 10), with typical daily values of β near 0.26 (bulrush site) and 0.13 (mixed site). This wholesale shift over to *LE* at the expense of *H* is largely a result of transpiration by the growing or mature vegetation, creating an obvious gulf between *H* and *LE* values in the energy-balance graphs (fig. 10). The tendency toward even lower β at the mixed site is somewhat unexpected, given the shallower water levels there, and suggests either greater partitioning toward *LE* by cattail and wocus transpiration at that site, or toward evaporation from areas of open shallow water associated with the wocus.

In early September, plants begin to senesce and turn brown, initiating a shift back toward equal partitioning between *H* and *LE*. During this time, *H* actually increases even though R_n rapidly decreases (fig. 10). During most of October, *H* is substantially greater than *LE* (β ≈ 2) primarily because October is dry (precipitation is discussed in Daily Evapotranspiration and Crop Coefficients) and above freezing. Therefore, what little precipitation falls is rainfall, which penetrates through the mostly vertical stalks to the lower stalk mat and the soil, where it is effectively decoupled from R_n and the overlying atmosphere. During November and December, the dead canopy lodges over, making a more supportive surface for the substantial snowfall during those months and providing a better exposure of the snow to R_n. As a result, β again approaches 1, which persists for the rest of the winter months.

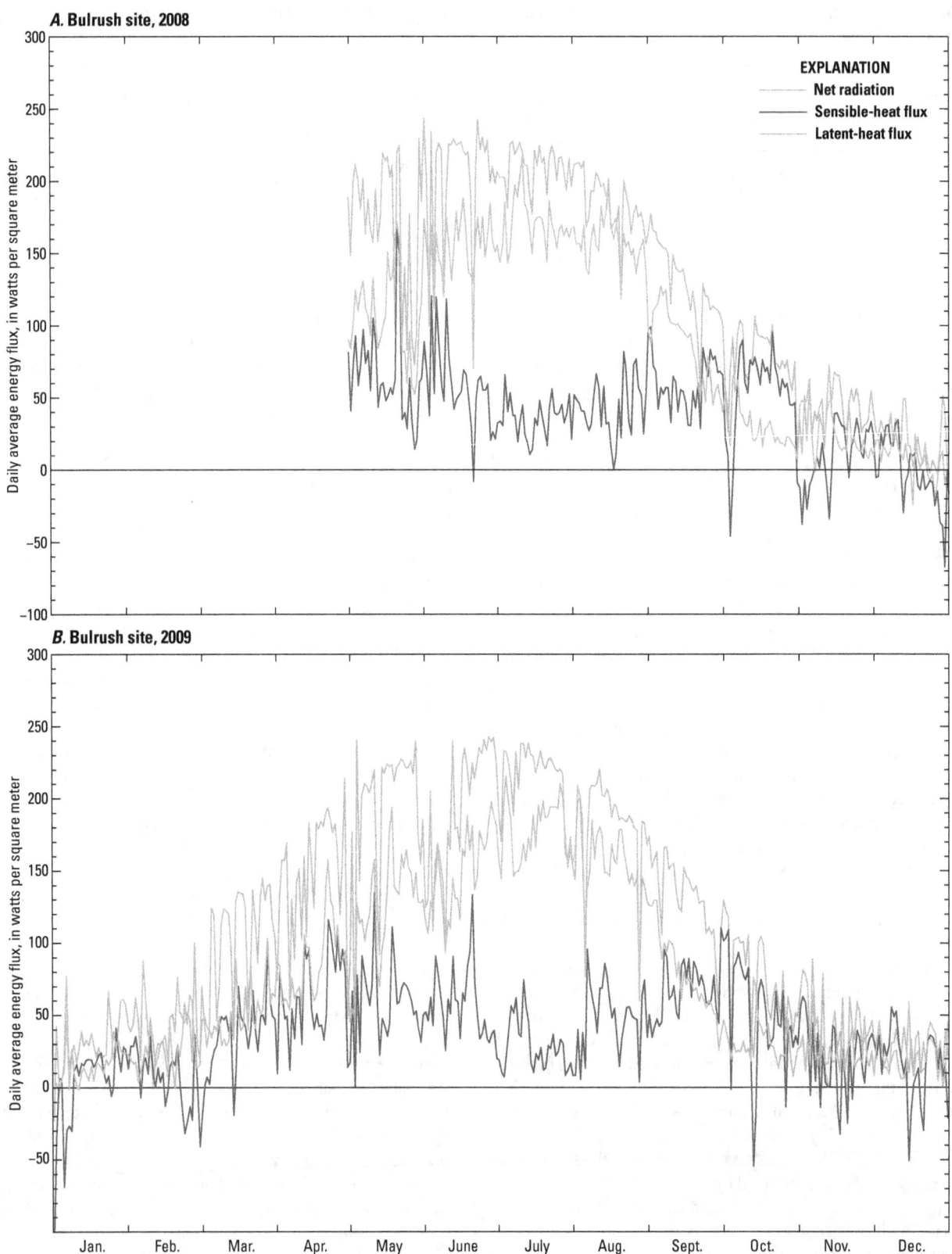

Figure 10.   Daily means of net radiation, sensible-heat flux, and latent-heat flux at bulrush site during *A*, 2008, *B*, 2009, and *C*, 2010, and at mixed site during *D*, 2008, *E*, 2009, and *F*, 2010.

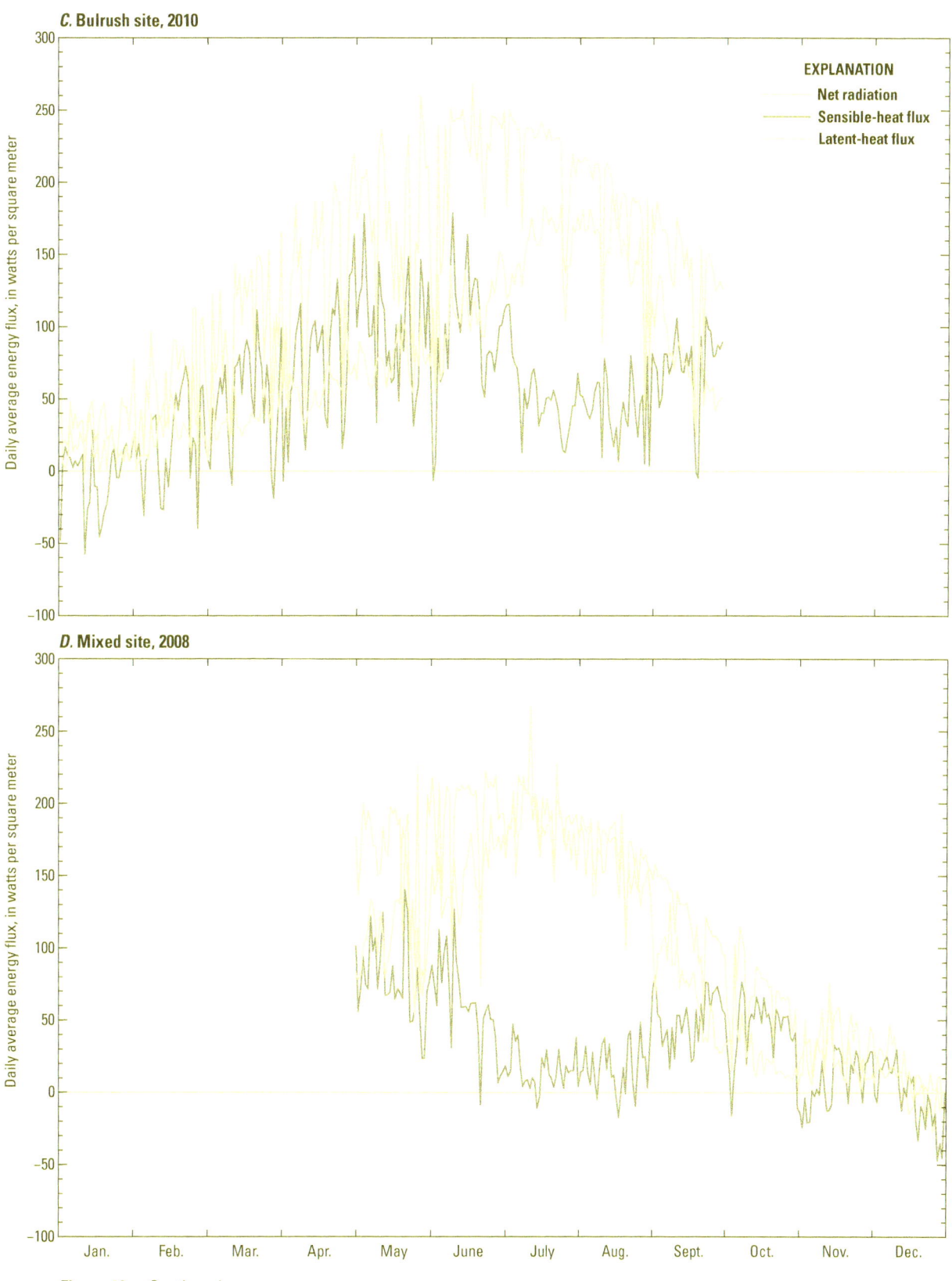

Figure 10.—Continued

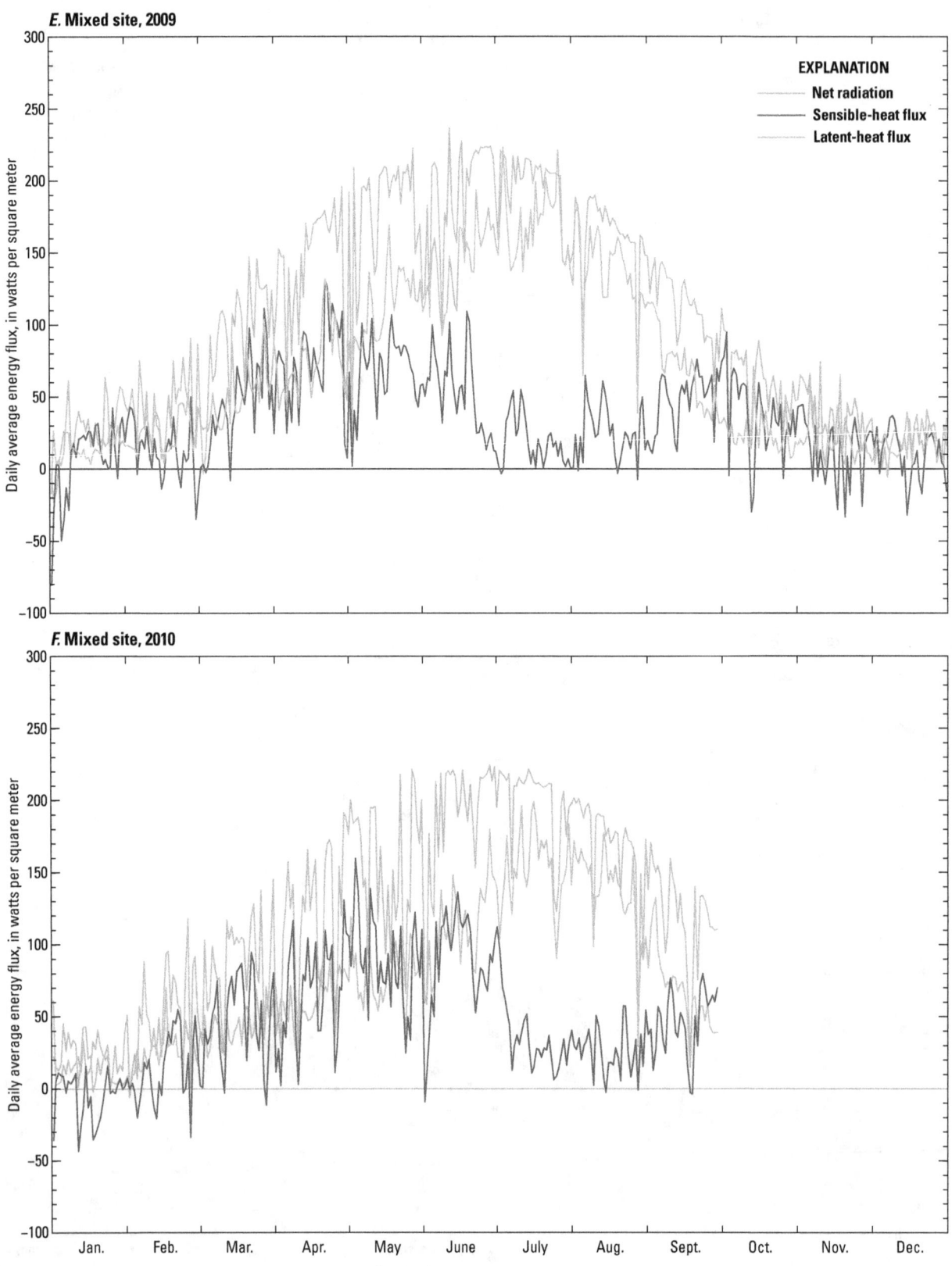

Figure 10.—Continued

Daily Evapotranspiration and Crop Coefficients

Daily values of ET and ET_r at both wetland sites during the study period (May 1, 2008, to September 29, 2010) are shown in figure 11. This study period brackets three growing seasons, from 2008 to 2010. A generally sinusoidal pattern in ET and ET_r can be seen during all 3 years, superimposed with daily variations above and below the mean pattern. The mean sinusoidal pattern is largely determined by the R_n input (fig. 10), modified somewhat by water level and vegetation properties, discussed in this section, below. The daily variations are also closely related to variations in R_n (fig. 10), and to precipitation, discussed in this section, below. A high degree of correlation can be seen between bulrush and mixed site ET, and between ET_r and ET at both sites. Both short term (daily) and longer term (weekly to monthly) variations in ET_r are reflected in the measured ET at both sites. On a daily basis, ET and ET_r alternate relative magnitudes to some degree, although on average $ET_r > ET$. Periods when ET is noticeably less than ET_r tend to be in the spring (May 2008; parts of April and May 2009; and parts of March, April, May, and June 2010) and fall (most of September and October 2008; late August through early October 2009; and parts of September 2010). During much of winter and late summer, ET is often indistinguishable from ET_r at the daily time step shown in figure 11.

Daily values of K_c at both wetland sites and precipitation (P) at the Klamath Falls weather station (KFLO) during the study period are shown in figure 12. The KFLO record is only an approximate indicator of timing and amount of P at the Upper Klamath NWR because the KFLO station is 44 km from the study sites. The most obvious feature in these plots is that daily K_c is generally predictable and well behaved from June through September, whereas K_c is quite variable (noisy) the rest of the year. The noisy periods consist of somewhat sustained intervals when K_c is generally small, punctuated with short-lived spikes in K_c, usually during and just after days of P (this relation between K_c and P is only approximate because of the distant KFLO location). This seasonal dependence of K_c variability is caused by (1) the seasonal change in the magnitude of ET_r; (2) the seasonal change in the amount and frequency of P; and (3) the seasonal change in sensitivity of canopy energy partitioning to inputs of P. During the growing season, larger ET_r provides a more stable denominator in the calculation of K_c ($K_c = ET/ET_r$), whereas measurement and modeling uncertainty in ET_r has a greater proportional effect on K_c when fluxes are small, during the non-growing season. Coincidentally, very little P falls between mid-June and the end of September, but P is instead heavily skewed toward the winter months. Therefore, the potential for K_c to spike in response to P intercepted by the canopy

occurs more frequently during the non-growing season. Lastly, interception has little effect on energy partitioning during the growing season, because (1) transpiration is already large, nearly satisfying the atmospheric demand; and (2) the live leaves are nearly vertical, shunting the water to lower canopy layers where it is less available for evaporation.

During the non-growing season, the bent over, more nearly horizontal leaves intercept more rain and snow and provide better exposure of the intercepted water to incoming radiation and better aerodynamic transport back to the atmosphere. This process shifts energy partitioning away from H toward LE, causing K_c to spike. For example, note the subdued response of K_c to P of 3 mm or more during the growing season on August 6, 2008, June 15 and 30, 2009, August 6-7, 2009, and July 25, 2010. In contrast, when the canopy is dormant, interception greatly enhances the ET rate, causing K_c to spike to values well above 1 for short periods, while the intercepted water evaporates (numerous examples in fig. 12). A reasonable correlation can be seen between the magnitude of daily P (measured at KFLO) and K_c during the non-growing season (fig. 12).

Daily values of K_c greater than about 2 require some explanation, considering the proximity of the study sites to the AGKO weather station (and the concomitant similarity in weather), and the high degree of water availability occurring at the study sites and assumed in the computation of ET_r using AGKO data. At the bulrush site, all days of $K_c > 2$ ($n=45$) occurred either before May 6 or after October 3. At the mixed site, all days of $K_c > 2$ ($n=32$) occurred either before May 11 or after November 5. These periods correspond to times before the emergence of new growth from the dead canopy litter layer in the spring, or after senescence in the fall (that is, roughly during the non-growing season). Probably the main cause of high K_c values is the assumption of a 45 s m^{-1} surface resistance in the computation of ET_r. During or just after precipitation, when the surface is wet, the actual surface resistance approaches zero. During the non-growing season, when the canopy is lodged over, the nearly horizontal leaf surfaces retain precipitation longer, sustaining high ET rates for one to a few days, depending on the amount of interception and the ensuing atmospheric demand. Once the canopy dries out, K_c typically returns to values well below one, reflecting dormancy and the lack of transpiration. This mechanism probably is compounded by a low signal-to-noise ratio in measured wetland ET and in computed ET_r, which are both relatively small at this time of year. Although K_c appears to be rather intractable during the non-growing season, its value is relatively unimportant because little ET occurs during this time. In addition, aggregation into biweekly periods reduces noise through averaging, as seen in the next section, Biweekly Evapotranspiration and Crop Coefficients.

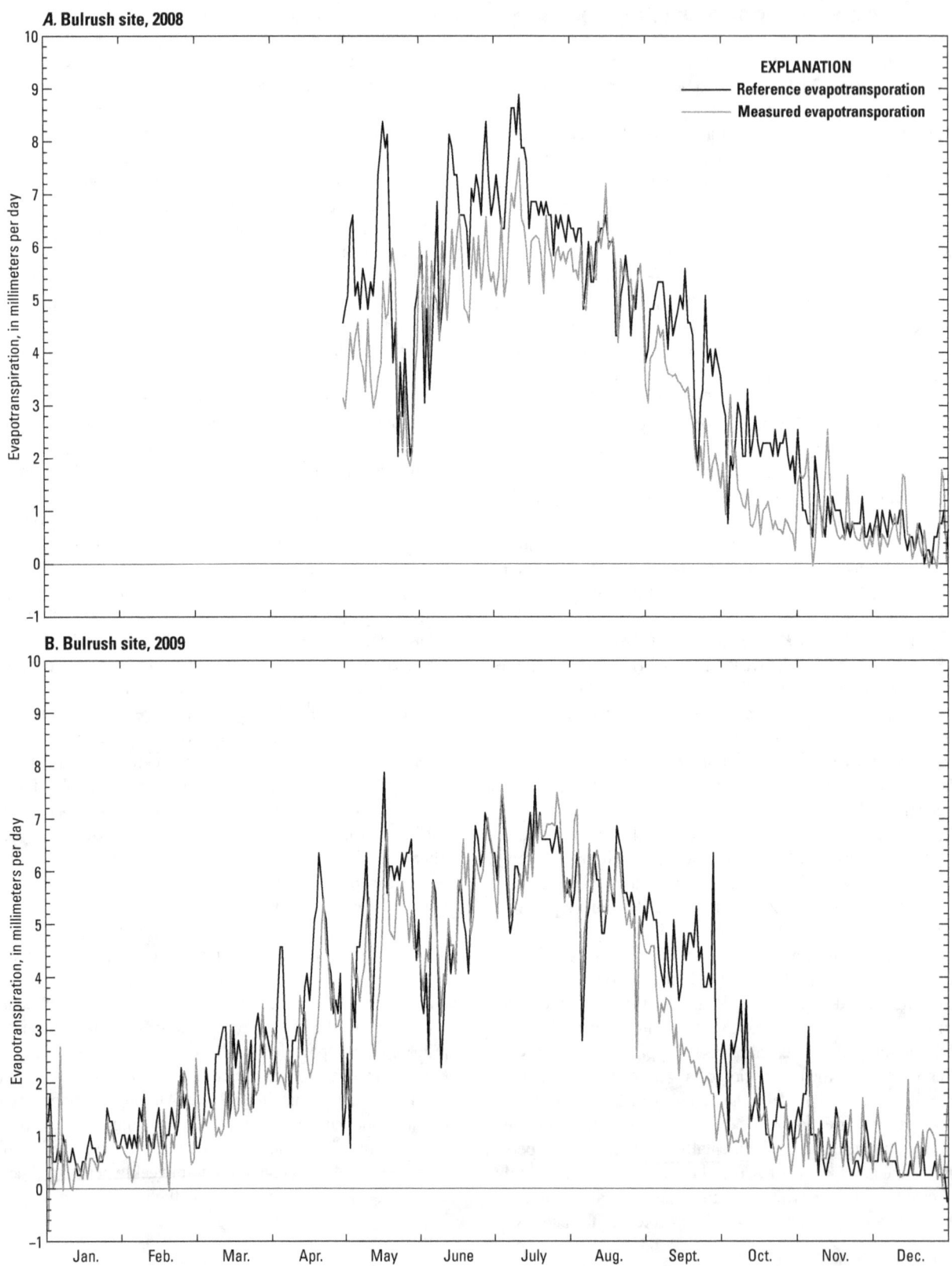

Figure 11.   Daily reference evapotranspiration and measured evapotranspiration at bulrush site during *A,* 2008, *B,* 2009, and *C,* 2010, and at mixed site during *D,* 2008, *E,* 2009, and *F,* 2010.

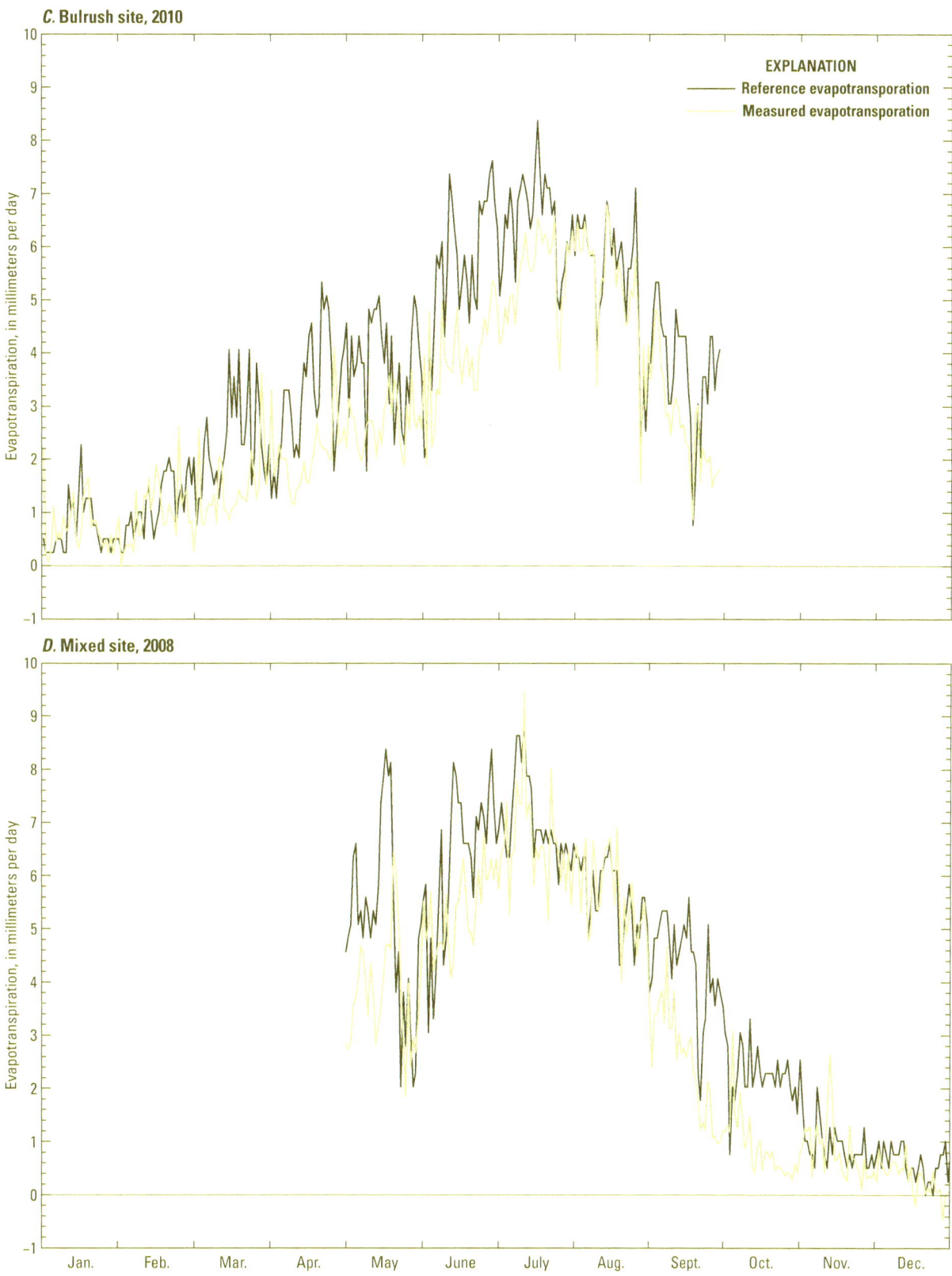

Figure 11.—Continued

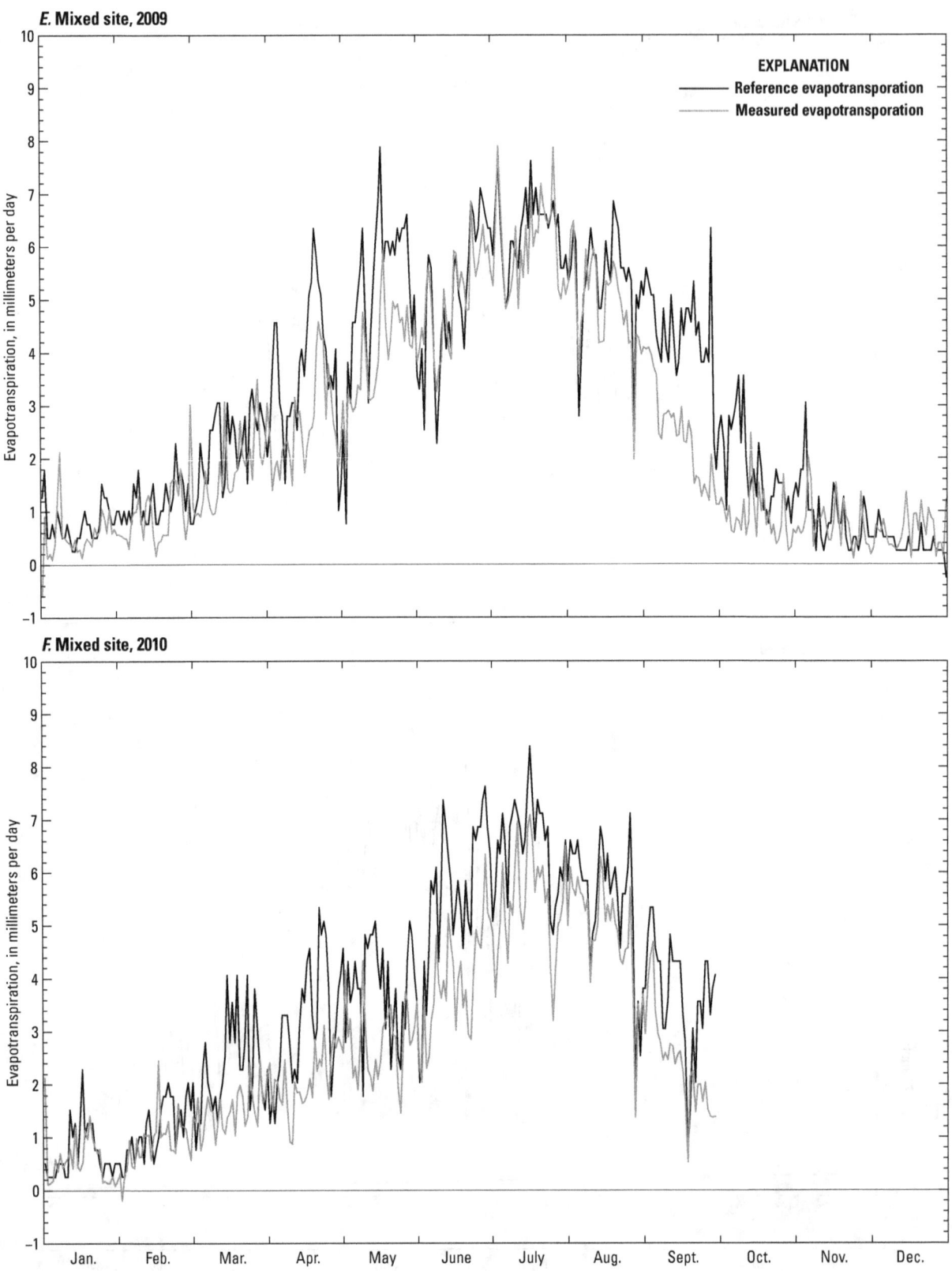

Figure 11.—Continued

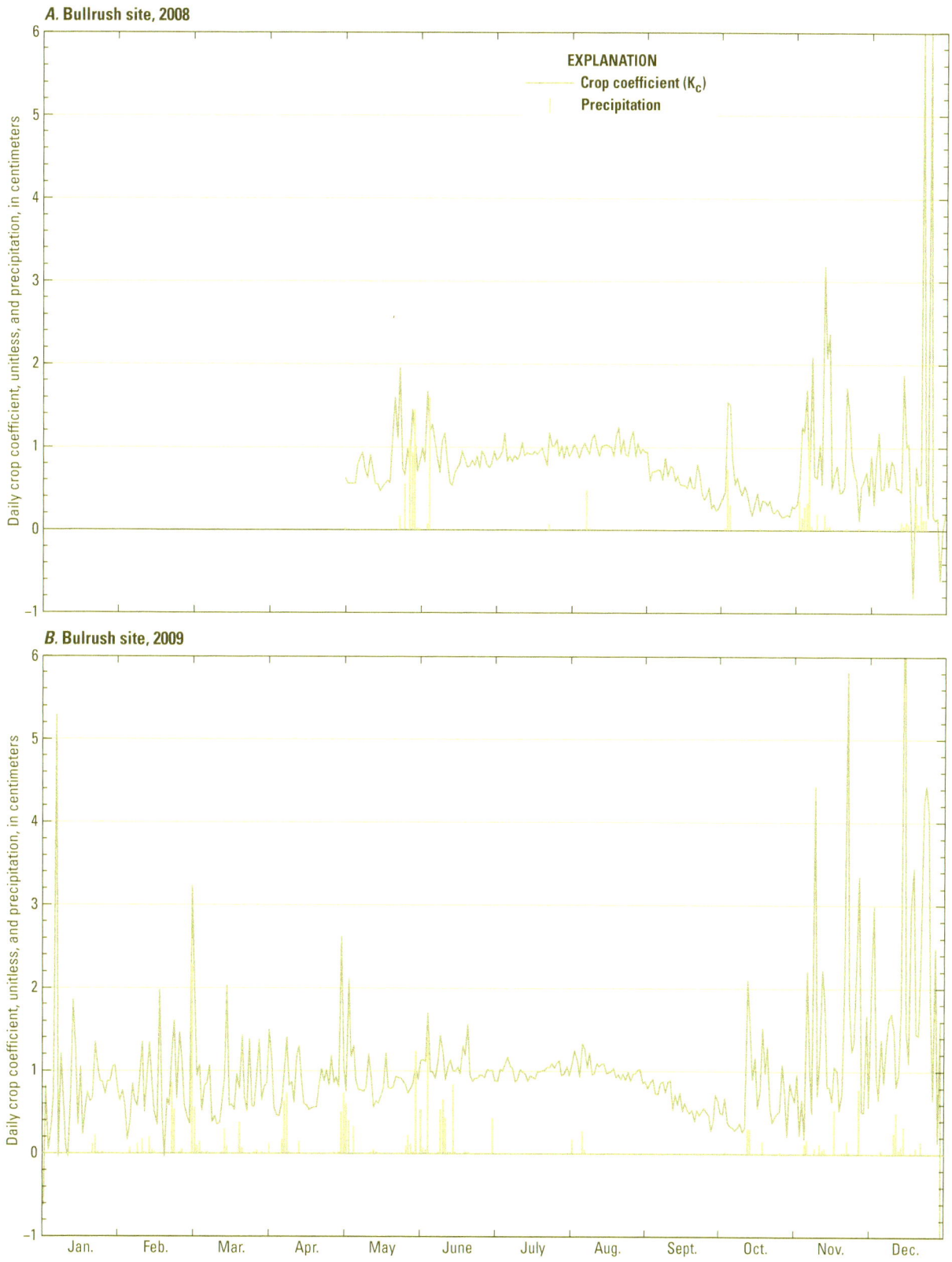

Figure 12. Daily crop coefficient at bulrush site during *A*, 2008, *B*, 2009, and *C*, 2010, and at mixed site during *D*, 2008, *E*, 2009, and *F*, 2010, and precipitation measured at Klamath Falls AgriMet station.

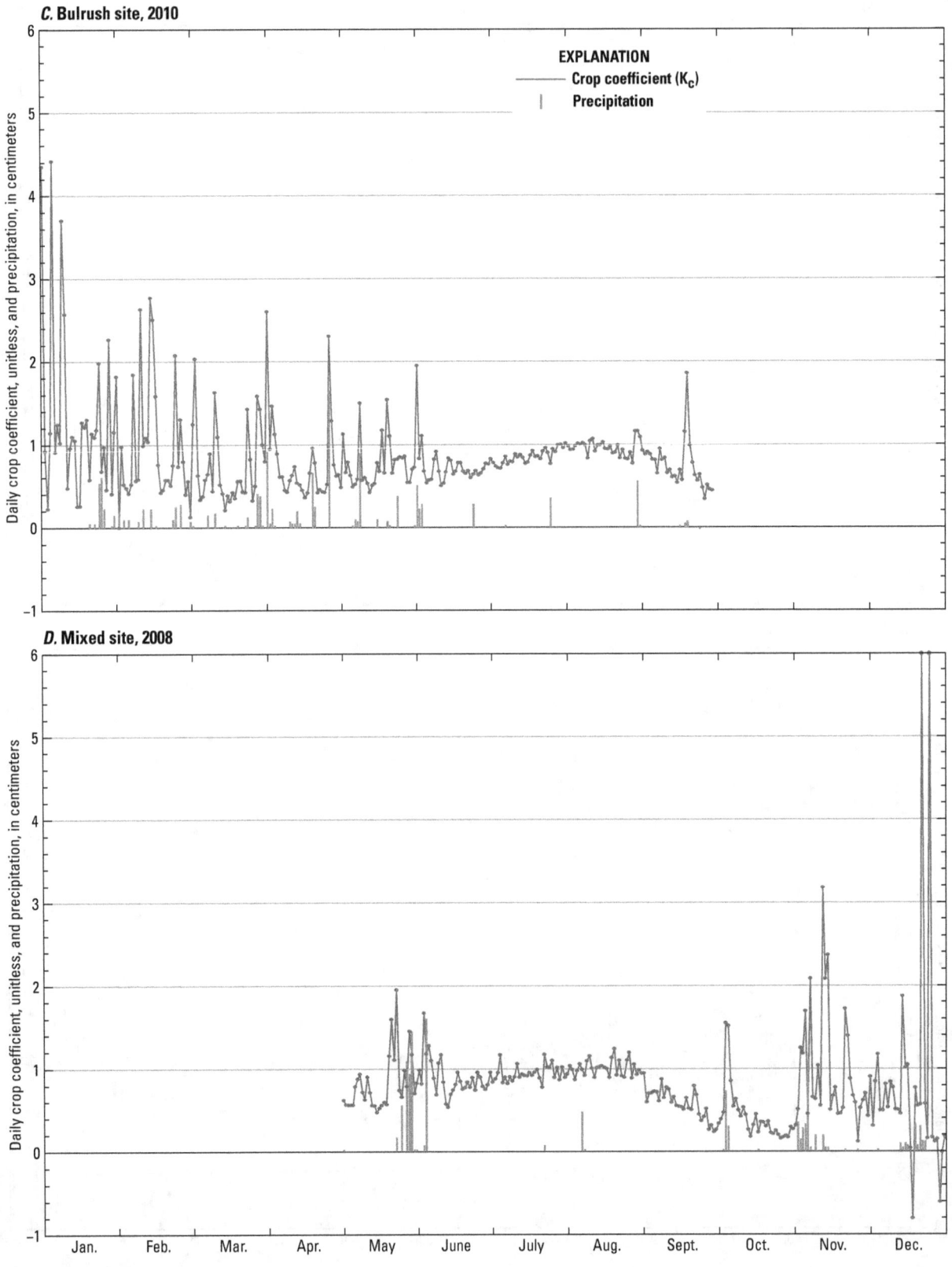

Figure 12.—Continued

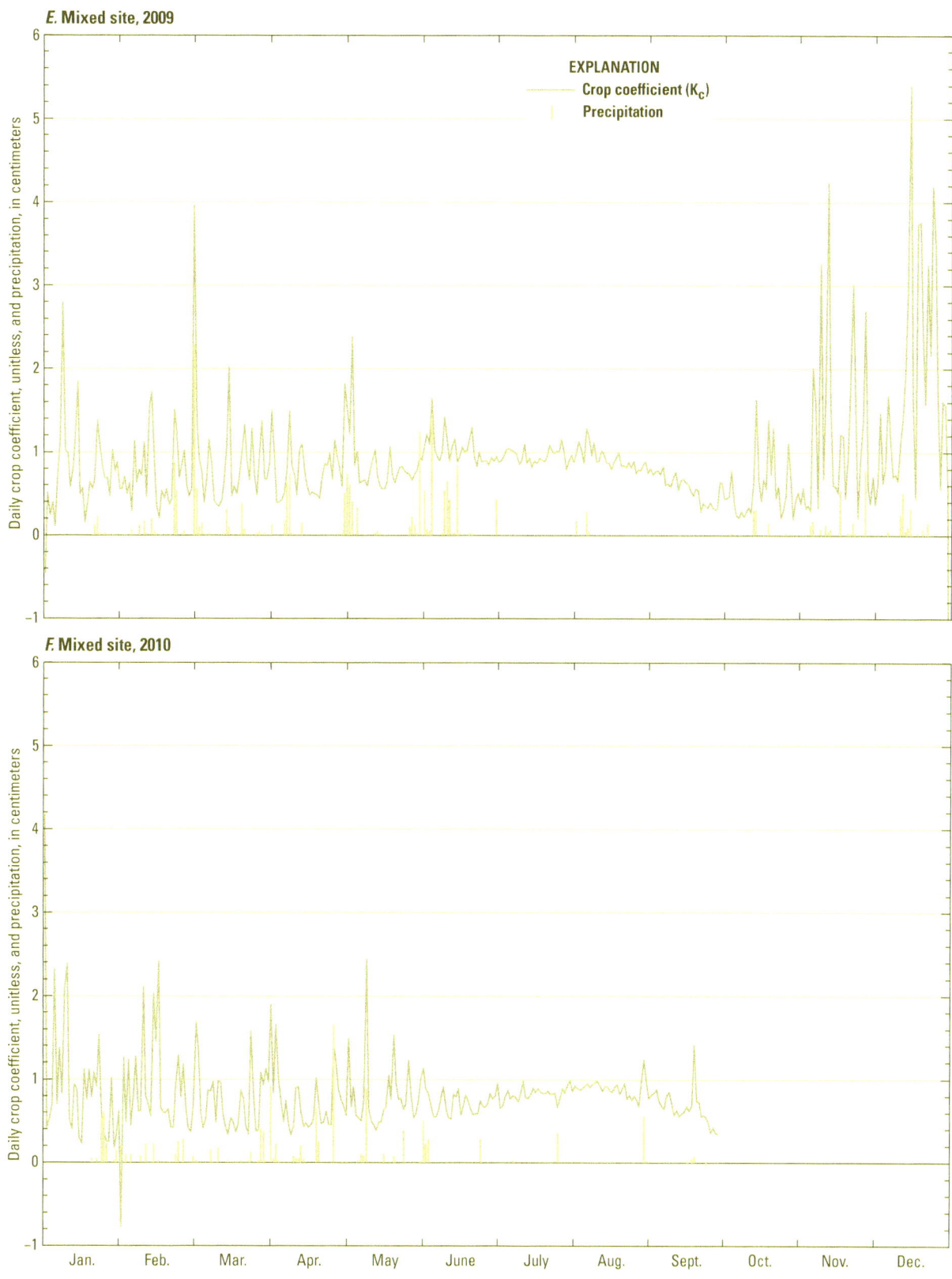

E. Mixed site, 2009

F. Mixed site, 2010

Figure 12.—Continued

An expanded view of daily K_c during the growing season of all 3 years is shown in figure 13. During the middle of the growing season, daily K_c follows a simple pattern rather consistently from year to year. From mid-June through mid-September, K_c is relatively well behaved and follows a concave-downward curve. Although the canopy begins growing in early May, K_c values in May and early June are still somewhat noisy because the new growth has not fully emerged from the dead stalk mat and interception causes spikes in K_c. During late September, canopy senescence has progressed sufficiently that interception again begins to cause spikes in K_c. A base value of K_c (analogous to base flow in a stream) can be established during these periods, using the lowest values as a guide to estimate dry-canopy K_c. Combining this base value with the mean behavior during the middle of the season leads to simple piecewise linear descriptions of K_c at both sites—as shown in figure 13 and quantified in table 5, that approximate the concave-downward curves. These piecewise linear approximations were determined by eye (visual examination). Based on spikes occurring from May 1 through about June 20, and after about September 15, a suggested value of $K_c = 1.5$ could be used on days of rainfall >3 mm during these periods (superseding the piecewise calculation), or on days following rainfall, if the rainfall occurs in the afternoon or evening.

This modeling scheme of daily K_c and ET was tested by assuming that all precipitation recorded at the KFLO AgriMet station occurred early in the day, triggering the specified value of $K_c = 1.5$ on the day of precipitation. A comparison of modeled and measured ET is shown in figure 14, and associated statistics are presented in table 6. These models are only marginally successful, as indicated by large scatter about the one-to-one line (fig. 14), small slope, r^2, and coefficient of efficiency (CE) values, and large intercept and $RMSE$ values (table 6). The CE is similar to r^2, except it ranges from 1 to -∞, is more rigorous, and requires equality as well as correlation between two variables to approach 1 (Nash

and Sutcliffe, 1970). Model performance may improve if the actual timing of precipitation were known, allowing the use of $K_c = 1.5$ on the day following precipitation when precipitation occurs late in the day. Further, if atmospheric demand on the day following precipitation is especially low (overcast conditions), the use of $K_c = 1.5$ could be extended to the next sunny day, possibly creating further improvement. However, model accuracy is limited by the 44-km distance between the KFLO precipitation measurement and the wetland sites (and concomitant decoupling of precipitation), and by the simplistic treatment of the complicated interactions between canopy structure, interception, and subsequent atmospheric demand and evaporation. While these models may provide an approximate estimate of daily ET at the wetland sites, much of the random variability can be removed by considering biweekly time steps, as in the following section.

Table 5.   Piecewise linear expressions of crop coefficient (K_c) at both study sites during the growing season.

[**Abbreviation**: DOY, day of year]

Site	Period	Equation
Bulrush	May 1–June 6 DOY 121–157	$K_c = 0.65$
	June 7–July 31 DOY 158–212	$K_c = 0.65 + 0.007407 \times (\text{DOY-158})$
	August 1–August 23 DOY 213–235	$K_c = 1.05$
	August 24–September 30 DOY 236–273	$K_c = 1.05 - 0.01622 \times (\text{DOY-236})$
Mixed	May 1–June 6 DOY 121–157	$K_c = 0.64$
	June 7–July 23 DOY 158–204	$K_c = 0.64 + 0.007609 \times (\text{DOY-158})$
	July 23–August 22 DOY 205–234	$K_c = 0.99$
	August 23–September 30 DOY 235–273	$K_c = 0.99 - 0.01605 \times (\text{DOY-235})$

Table 6.   Results of modeling daily evapotranspiration (ET) during the wetland growing season (May 1–September 30, 2008–2010, $n = 457$) at both sites using piecewise linear models of K_c, superseded by setting $K_c = 1.5$ during days of precipitation greater than 3 millimeters.

[Slope and intercept are of ordinary least squares best-fit line between modeled and measured ET. **Abbreviations**: $\overline{ET}$, mean evapotranspiration during the 2008–2010 growing seasons; mea, measured; mod, modeled; r^2, coefficient of determination; CE, coefficient of efficiency; $RMSE$, root-mean-square-error; mm d^{-1}, millimeters per day]

Site	ET_{mea} (mm d^{-1})	ET_{mod} (mm d^{-1})	r^2	CE	Slope	Intercept (mm d^{-1})	$RMSE$ (mm d^{-1})
Bulrush	4.537	4.499	0.710	0.652	0.779	1.033	0.902
Mixed	4.376	4.340	0.733	0.704	0.834	0.756	0.870

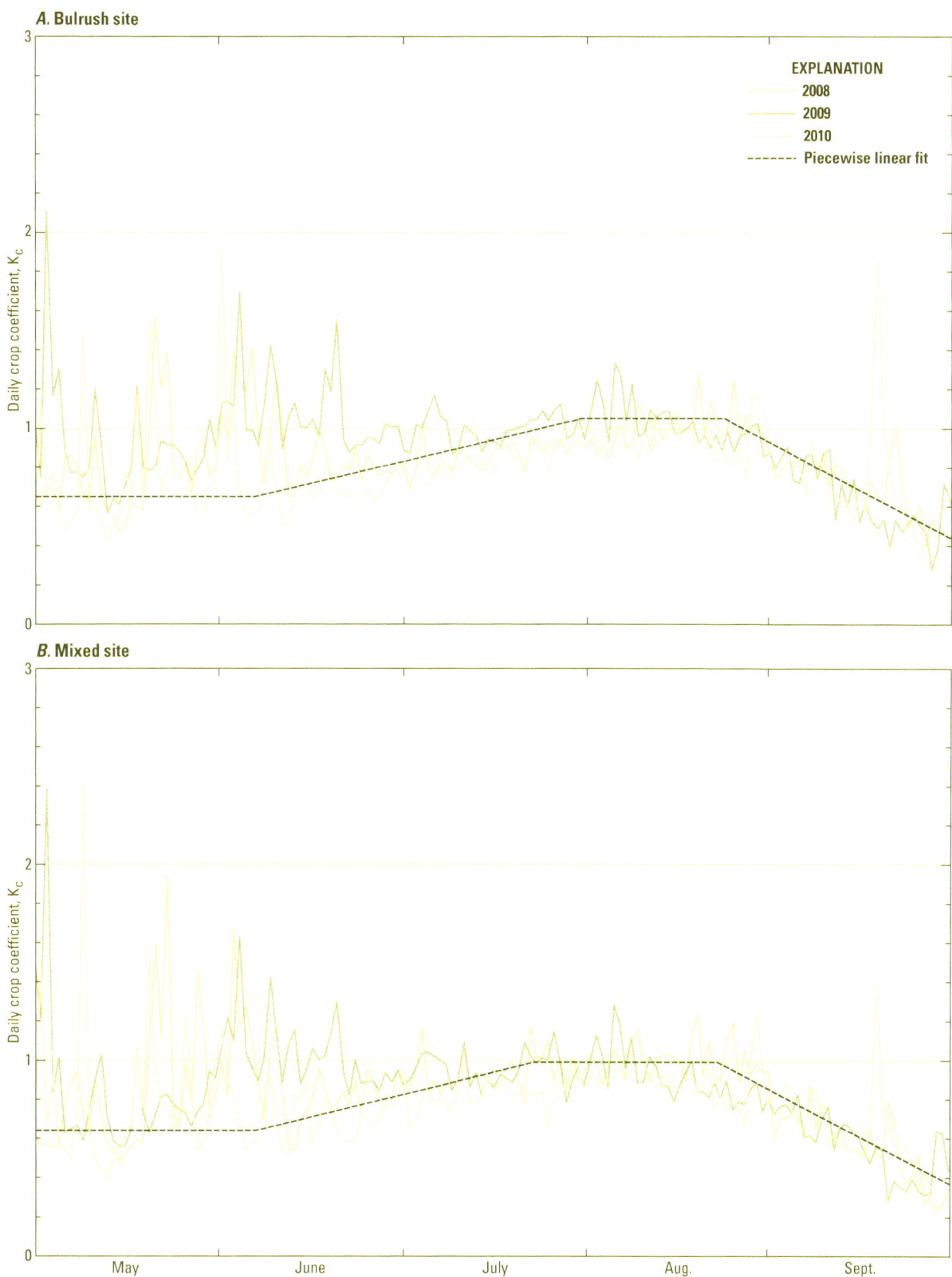

Figure 13. Detailed graph of daily crop coefficient during the growing seasons of all 3 years at *A*, the bulrush site, and *B*, the mixed site, and piecewise linear fits to the means from about June 20 to September 15, and to base values outside of this period. Piecewise linear fits determined by eye.

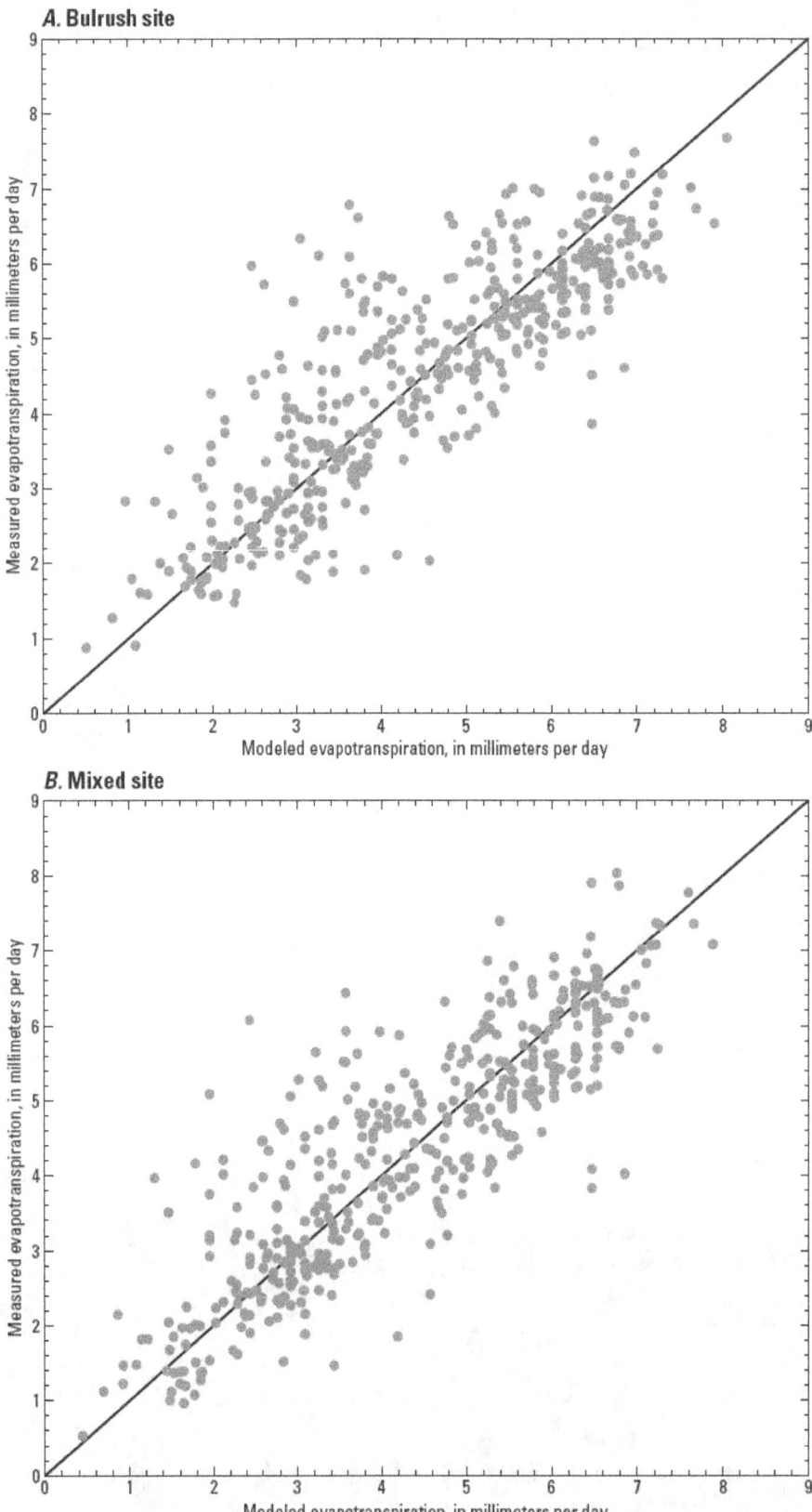

Figure 14.   Comparison of daily evapotranspiration (*ET*) measured during the growing season (May 1–September 30) of 2008 through 2010 with *ET* modeled using the piecewise linear approach for the *A*, bulrush site and *B*, mixed site.

The rising trend in daily K_c during April through mid-July is noticeably delayed at both sites in 2010, a year of substantially lower water levels (fig. 8). During this period in 2010, ET is also noticeably smaller than in 2008 and 2009 (fig. 11). The lower K_c values in early 2010 probably were not caused by canopy stress because water levels were at or above land surface by late February at the bulrush site and by early April at the mixed site, well before new growth began. The major meteorological variables affecting ET were averaged for the period April 1–June 30, and a comparison of these averages for 2009 and 2010 is given in table 7. Wind speed is not included because its effect on ET is small and variable depending on other environmental conditions (Campbell and Norman, 1998). Spring 2010 was more overcast, humid, and cool, but less rainy than spring 2009. These variables all contributed to lower ET in 2010, but the reduced solar radiation, vapor pressure deficit, and temperature do not account for the small K_c because they affect ET_r to about the same proportion as they affect ET. The similar reduction in ET_r and ET from these environmental conditions can be seen by comparing figures 11B, C, E, and F. The reduced rainfall in 2010 probably did contribute to smaller K_c through reduced evaporation of intercepted rainfall, but as discussed earlier in this section, the effect of rainfall is short-lived, causing upward spikes in K_c, and even the base values of K_c (during dry periods between rainfall) are substantially lower in 2010 than in 2009 and 2008 (fig. 12). Apparently some other factor contributed to the reduction in ET during spring 2010 without a corresponding reduction in ET_r, leading to consistently smaller values of K_c.

A mechanism for a relation between K_c and standing water level has been proposed by German (2000), who measured ET at multiple sites in the Florida Everglades—vegetated with sawgrass, spike rush, muhly grass, and cattail; vegetation similar to that at the present study site. If canopy biomass is distributed evenly in the vertical direction, light penetration from above decreases with depth into the canopy approximately according to Beer's law. Conversely, the amount of light reaching an underlying water surface increases as water-surface elevation increases. At the Everglades site, the density of plant material (primarily the dead stalk understory) was greatest near the land surface and decreased with distance above land surface (Carter and others, 1999). German reasoned that this canopy architecture would enhance the extinction effect compared to that of an evenly distributed canopy, greatly increasing the proportion of water surface receiving solar radiation at higher water levels. Greater radiation input to the water surface causes greater partitioning to LE. At lower water levels, a greater proportion of dead plant material per unit horizontal area is exposed to sunlight, generating greater H at the expense of LE. We expand on German's (2000) observation by noting that lower water levels also increase the aerodynamic resistance from the water surface to the free atmosphere, further shifting the energy exchange away from the water surface, toward the upper, dead canopy. At the current study site, canopy architecture is similar to that documented in the Everglades, and although detailed measurements were not made, many photographs taken during site visits substantiate the greater density of dead plant material near the ground. This dependence of energy partitioning on water level very likely occurs at the present study site, and the unusually small value of K_c during spring 2010 probably was at least partially related to the unusually low water levels during that time.

Table 7.   Comparison of environmental variables affecting evapotranspiration from April 1 to June 30, 2009 and 2010, at both wetland sites.

[All values shown are period means, except rainfall, which is period total; Ratio, ratio of 2010 value to 2009 value; Diff., 2009 value minus 2010 value. **Abbreviations**: kPa, kilopascals; mm, millimeter; W m^{-2}, watts per square meter; C, degrees Celsius]

Site	Solar radiation (W m^{-2})			Vapor pressure deficit (kPa)			Air temperature (°C)			Rainfall (mm)		
	2009	2010	Ratio	2009	2010	Ratio	2009	2010	Diff.	2009	2010	Ratio
Bulrush	289	274	0.95	0.624	0.517	0.83	10.81	8.17	2.64	249	168	0.68
Mixed	284	263	0.93	0.621	0.500	0.81	10.67	8.10	2.57	289	196	0.68

Biweekly Evapotranspiration and Crop Coefficients

Biweekly ET_r and measured ET at both sites are shown by year in figure 15. A reasonable correlation can be seen between the two sites and between measured ET and ET_r. On average, ET_r exceeds ET, although some exceptions can be seen. Bulrush site ET tends to exceed mixed site ET, also with notable exceptions (for example, midsummer, 2008). Although the timing of the relation between ET and ET_r changes from year to year, ET tends to approach ET_r most reliably from mid-July to mid-August. During the winter, when fluxes are small, ET approaches and even substantially exceeds ET_r, but no consistent patterns are apparent. In December 2009, ET at both sites exceeds ET_r by more than a factor of 2.

Biweekly K_c values were computed from the biweekly ET and ET_r rates, and the K_c values are shown in figure 16, along with biweekly precipitation (P) totals. As with daily K_c, values are somewhat predictable during the growing season (May through September) and are quite variable during other times of the year. Values of daily K_c that spike to 5 or 6 in November through January (fig. 12) are somewhat moderated by aggregation into biweekly periods, but values still exceed 2 in December 2009 due to the small values of ET_r during that time. On a biweekly basis, K_c is less correlated with P than on a daily basis (fig. 16). For example, relatively large P in early June 2009 and mid-April 2010 did not result in large values of K_c; in fact, K_c was smaller in December 2008 than in December 2009—although P was greater in 2008. This lack of correspondence between P and K_c probably is related to: (1) the timing and frequency of P; (2) the tendency for P to be sequestered by the lower canopy understory and soil; and (3) the abundant soil water content during the growing season. During the non-growing season, rainfall amounts that exceed the interception capacity of the dormant canopy drain to lower layers and the soil, where evaporation is reduced due to shading and decreased aerodynamic transport. A greater percentage of drainage occurs from a few large rainfalls than from many small rainfalls of equal total depth, reducing the percentage available for evaporation. This mechanism reduces the correlation between biweekly K_c and P, instead making K_c more dependent on the number of rainfall events during the period. Snowfall tends to remain more elevated in the canopy than rainfall, but wind redistributes the snow to lower levels, and solar radiation melts the snow on warmer days, causing some drainage and obscuring the linkage between K_c and P. During the growing season, evaporation of interception is comparable to transpiration due to the ample soil moisture, reducing the impact of interception on ET. While interception of P causes noticeable spikes in daily K_c, this effect becomes masked by other processes on a biweekly basis. Therefore, the biweekly magnitude of P does not appear to be useful in predicting the biweekly value of K_c.

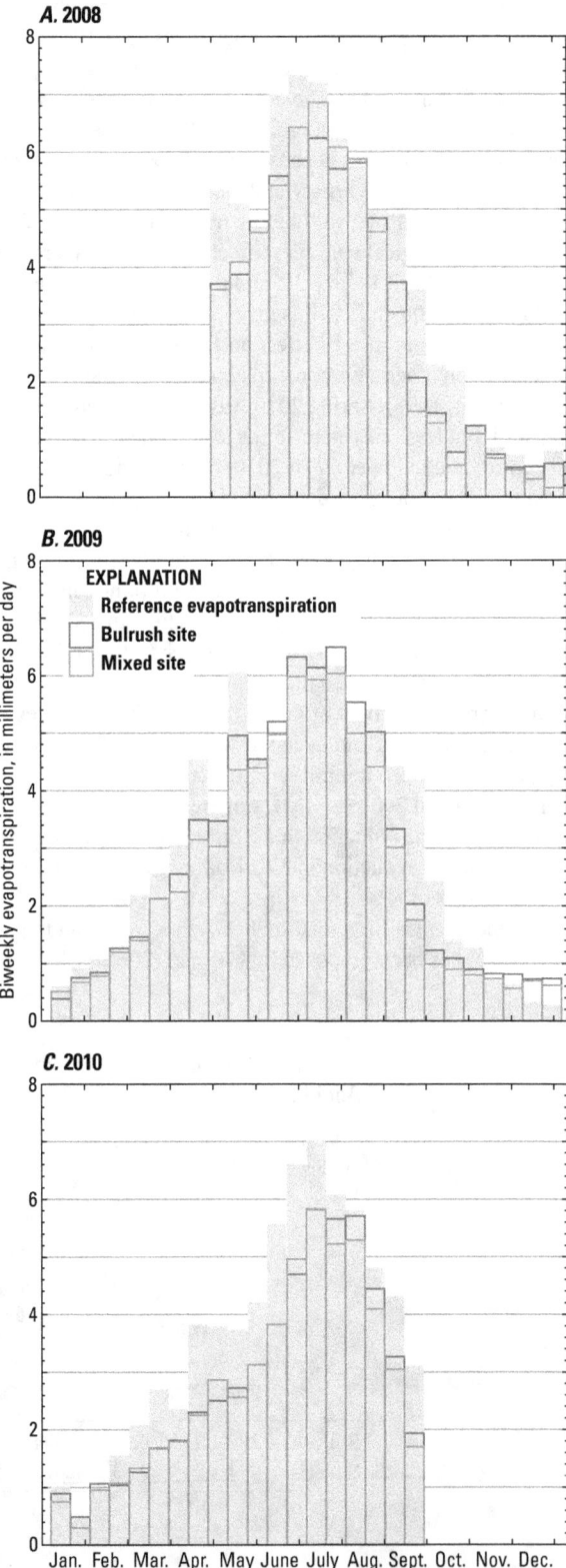

Figure 15. Biweekly reference evapotranspiration from Agency Lake AgriMet station and measured evapotranspiration at bulrush and mixed sites during *A*, 2008, *B*, 2009, and *C*, 2010.

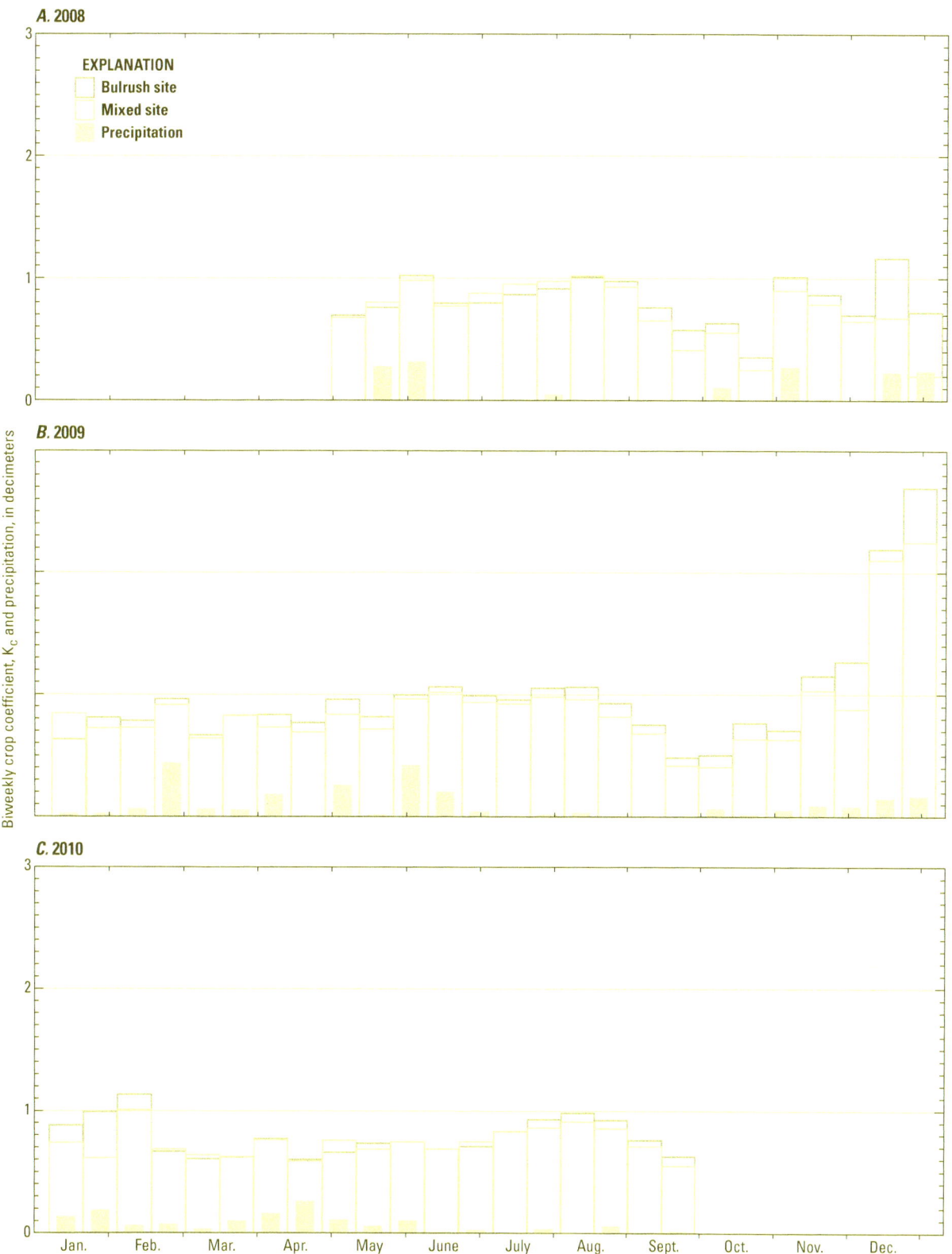

Figure 16. Biweekly crop coefficients at bulrush and mixed sites, and precipitation at Klamath Falls AgriMet station during *A*, 2008, *B*, 2009, and *C*, 2010.

Ensemble averages of ET and ET_r were computed using data from all 3 years and were parsed into 26 biweekly periods to create a 1-year average record. Ensemble average K_c for each biweekly period was then calculated as the ratio of average ET to average ET_r. By averaging fluxes from the corresponding biweekly periods each year, the non-linear effects of averaging unusually large or small K_c values are removed. Due to the start (May 1, 2008) and end (September 29, 2010) of the study period, averages are computed from 3 years during the growing season (May through September), and 2 years otherwise. The resulting time series of ensemble average biweekly K_c is reasonably predictable during the growing season and somewhat more erratic otherwise (fig. 17). Values of ensemble average K_c during the growing season are listed in table 8.

During the growing season, ensemble average K_c rises almost monotonically from a value near 0.75 on May 1, to a peak near 1.0 in mid-August, then declines monotonically to a value near 0.5 by late September. The only substantial perturbation in this pattern is during late May-early June, when K_c is greater than the subsequent 2 or 3 values. This boost in ET at both sites may be related to fortuitous timing and amounts of rainfall, primarily in 2008 and 2009 (fig. 12). On the rising limb of K_c, values at the two sites are nearly equal, with the bulrush K_c slightly greater than the mixed site K_c on average. On the descending limb, the bulrush K_c is substantially and consistently greater than the mixed site K_c. The tabled values of ensemble average K_c are used to model biweekly ET during the 3 growing seasons, as discussed later in this section.

During the non-growing season, ensemble average K_c varies somewhat unpredictably; large variations occur between adjacent periods and between sites (fig. 17). In particular, the very large (>2) values in late December 2009 (fig. 16B) have been somewhat moderated by the 2008 data, but the 2-year average still stands out from the rest, especially at the bulrush site. In addition, K_c values in October are unusually small compared to values during the rest of the non-growing season. During the study, October precipitation at the Klamath Falls KFLO station was 52 percent of the average for 1999 through 2011. Part of the random variability in K_c is probably related to the small sample size ($n=2$ years) during the non-growing season, combined with the small ET_r values as described in Daily Evapotranspiration and Crop Coefficients. It is therefore suggested that the extreme values of K_c (mostly during October and December) are not representative of the long-term, and that a single mean value of K_c can adequately characterize the whole non-growing season. Consequently,

Table 8.   Ensemble mean crop coefficients (K_c) during growing season biweekly periods and mean non-growing season K_c values at both study sites.

Period	Bulrush	Mixed
April 30–May 13	0.760	0.745
May 14–May 27	0.775	0.739
May 28–June 10	0.926	0.900
June 11–June 24	0.836	0.814
June 25–July 8	0.829	0.853
July 9–July 22	0.883	0.903
July 23–August 5	0.965	0.938
August 6–August 19	1.017	0.964
August 20–September 2	0.942	0.864
September 3–September 16	0.757	0.678
September 17–September 30	0.555	0.453
Non-growing season	0.758	0.683

ET and ET_r were averaged from October through April, and a single non-growing season K_c was computed for each site. These values are 0.758 and 0.683 at the bulrush and mixed sites, respectively, and are listed in table 8.

Ensemble average values of K_c from the growing season (fig. 17; table 8) and mean values from the non-growing season (table 8) were multiplied by biweekly values of ET_r to test the adequacy of the average K_c values to reproduce the measured ET values at both sites during the entire study, consisting of 63 biweekly periods. The results of this comparison are shown in figure 18, and associated statistics are presented in table 9. Overall, performance of the average K_c approach is very good, with r^2 values greater than 0.96, $RMSE$ values less than 0.4 mm d^{-1}, and slopes and intercepts of best-fit lines near 1 and 0, respectively. This is not the most stringent test of the method, because the seasonal K_c values were determined from the test data rather than from an independent data set. However, a fair amount of variability in K_c occurred from year to year (during the same biweekly period) at both sites (fig. 16), and the ability of the model to reproduce the measured ET is still quite good. For example, water levels were about 0.5 to 0.2 m lower in 2010 than in 2008 and 2009 (fig. 8) during the first half of the year. As a result of this or other factors, K_c values were noticeably lower during the early 2010 growing season (fig. 16), and inclusion of these data lends some generality to the K_c model by averaging in the unusual year. The resulting model only slightly overestimates ET in 2010 (the red squares in fig. 18), while still maintaining relatively good accuracy overall.

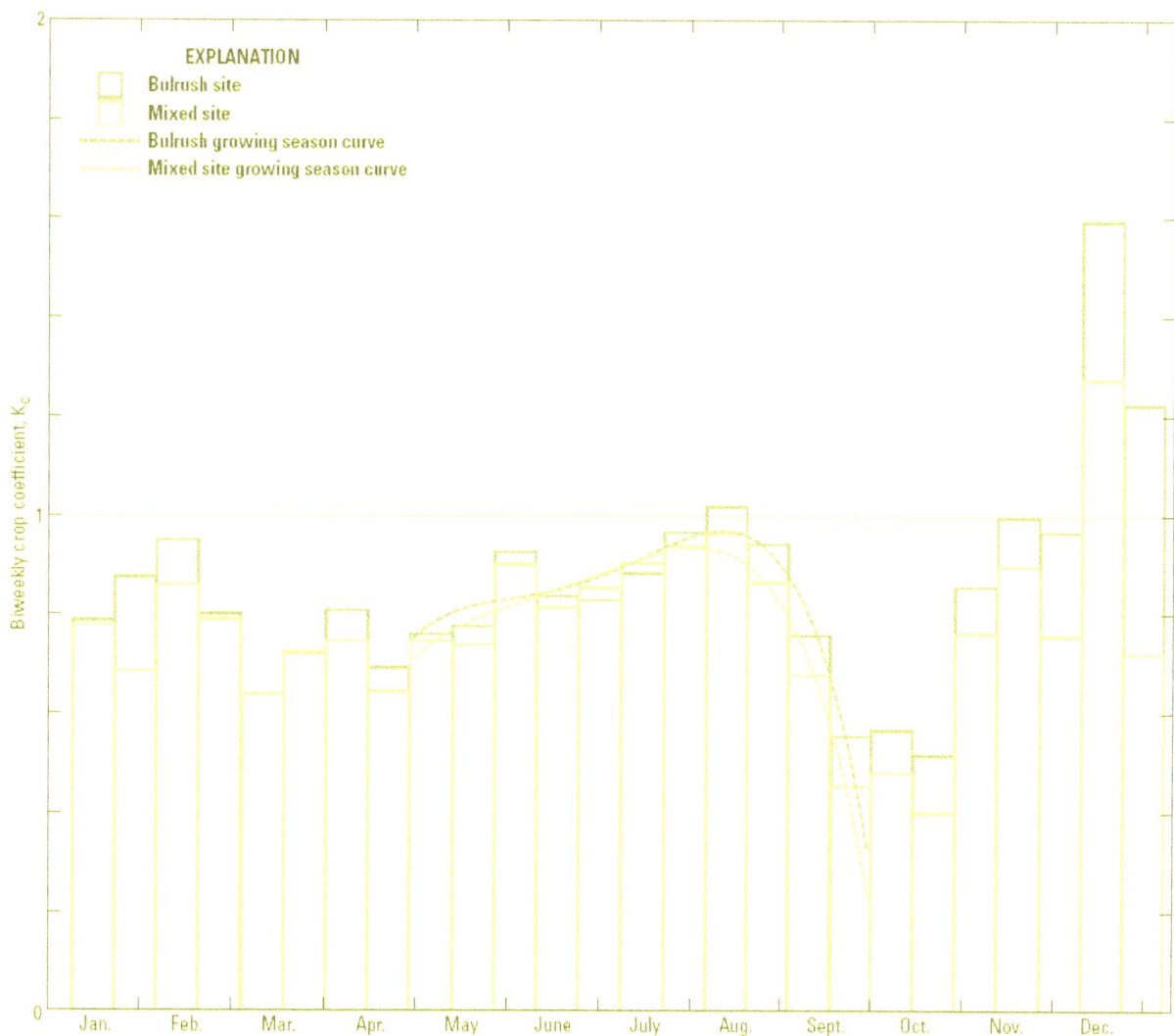

Figure 17. Biweekly ensemble average crop coefficients at bulrush and mixed sites, and fourth-order polynomial fits to the crop coefficients during the growing season (May–September).

Table 9. Results of modeling biweekly evapotranspiration (ET) during the study period (n = 63) at both sites using the ensemble-mean K_c approach during the growing season and the mean K_c approach during the non-growing season.

[Slope and intercept are of ordinary least squares best-fit line between modeled and measured ET. **Abbreviations:** CE, coefficient of efficiency; $\overline{ET}$, mean ET during the study period; mea, measured; mod, modeled; mm d^{-1}, millimeters per day; r^2, coefficient of determination; $RMSE$, root-mean-square-error]

Site	$\overline{ET}_{mea}$ (mm d^{-1})	$\overline{ET}_{mod}$ (mm d^{-1})	r^2	CE	Slope	Intercept (mm d^{-1})	RMSE (mm d^{-1})
Bulrush	2.926	2.926	0.962	0.961	0.977	0.067	0.395
Mixed	2.787	2.787	0.971	0.970	0.986	0.039	0.347

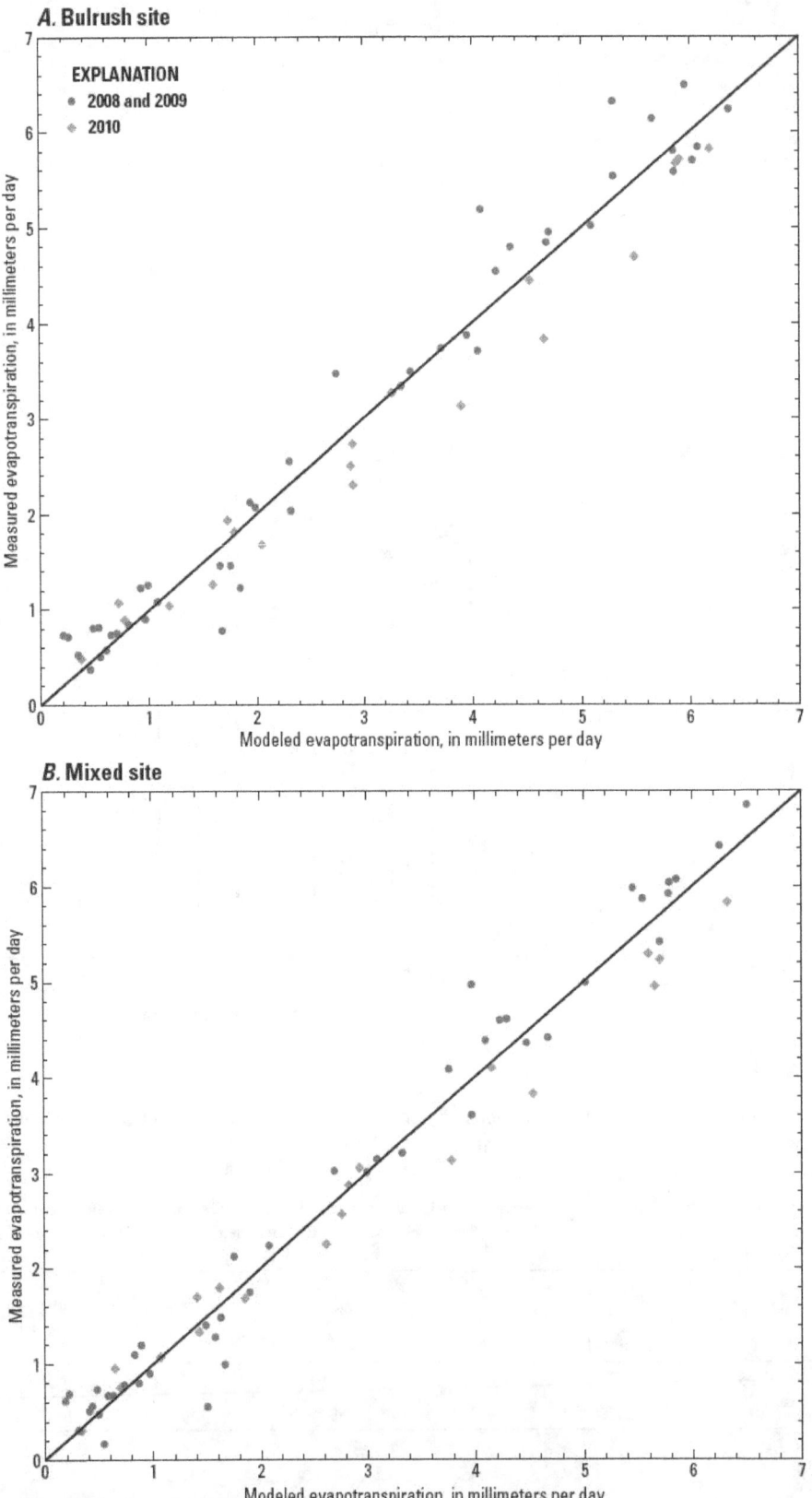

Figure 18. Comparison of biweekly evapotranspiration (ET) measured during the entire study period with ET modeled using the ensemble average K_c approach for the A, bulrush site and B, mixed site.

As an alternative to the discrete values of biweekly growing season ensemble average K_c in table 8, fourth-order polynomials were fit to the biweekly values and the resulting curves are shown in figure 17. Information for computing K_c in this manner is given in table 10. This approach provides a more generalized and automated computation of growing season K_c, but undoubtedly sacrifices accuracy in modeling *ET* during the study period compared to the K_c table approach (fig. 18; table 9). However, over periods of many years, this approach may perform equivalently to the K_c table approach.

Study-Period and Annual Evapotranspiration

On average during the study period, R_n at the bulrush and mixed sites was 126.4 W m^{-2} and 113.9 W m^{-2}, respectively. The smaller mixed site R_n of about 10 percent probably is related to differences in cloudiness, water depth, or vegetation type. The 4-component measurements of R_n indicate that this 10 percent difference is partitioned as follows: 3.8 percent to incoming solar, 4.3 percent to reflected solar, 1.5 percent to outgoing long-wave, and 0.4 percent to incoming long-wave. The smaller incoming solar at the mixed site is likely associated with the predominant south-westerly winds at that site, issuing from the Cascade Range through the mouth of Fourmile Creek at Pelican Bay (fig. 1), entraining montane cloudiness preferentially over the mixed site. The bulrush site is more removed from the mountains and from the Fourmile Creek air stream, as indicated by its more evenly distributed wind directions, and the site should experience sunnier skies as a result. Mean reflectivity of the bulrush site during the study was 0.128, compared to 0.156 at the mixed site. This

greater retention of solar radiation at the bulrush site was the largest factor in the R_n difference and is consistent with the deeper water at that site (Sumner and others, 2011). But the R_n difference also could be related to differences in vegetation color. Deeper water at the bulrush site also would tend to keep the surface cooler and account for the reduced outgoing long-wave radiation at that site. Although the mixed site was located among large patches of open water associated with wocus, wocus patches were distant enough from the station that they did not affect the measured R_n, but probably did affect the measured turbulent fluxes.

Average *ET* during the study period, during each year of study, and during a 3-year period was computed for both sites and is presented along with ET_r in table 11. The 2008 and 2010 annual values were computed by filling in the missing parts of 2008 (January 1–April 30) and 2010 (September 30–December 31) with data from the equivalent times in 2009.

Table 10. Coefficients of fourth-order polynomial fits to biweekly growing season ensemble average K_c values.

[Polynomials are of the form: $K_c = C0+C1(X)+C2(X^2)+C3(X^3)+C4(X^4)$, where $X = DOY/100$. Because of the powers of X involved, all significant digits shown are needed to get accurate results. This number of significant digits did not require use of double precision in FORTRAN. **Abbreviation:** *DOY*, day of year]

Site	Coefficient				
	C0	C1	C2	C3	C4
Bulrush	-9.411	23.9880	-20.9950	8.09919	-1.15345
Mixed	-6.712	17.2984	-15.1276	5.92317	-0.865068

Table 11. Mean evapotranspiration at the study sites and reference evapotranspiration at the Agency Lake (AGKO) weather station during the study period (May 1, 2008–September 29, 2010), each year of study, and a full 3-year period (January 1, 2008–December 31 2010).

[Full years of record in 2008 and 2010 were synthesized by using data from equivalent periods in 2009 to fill in missing periods in 2008 (January 1–April 30) and 2010 (September 30–December 31). The 3-year period is the mean of 2008, 2009, and 2010. **Abbreviations:** mm d^{-1}, millimeters per day; m yr^{-1}, meters per year]

Site	Study period		2008		2009		2010		3-year period	
	(mm d^{-1})	(m yr^{-1})	(mm d^{-1})	(m yr^{-1})	(mm d^{-1})	(m yr^{-1})	(mm d^{-1})	(m yr^{-1})	(mm d^{-1})	(m yr^{-1})
Bulrush	2.93	1.069	2.72	0.995	2.77	1.013	2.33	0.850	2.61	0.953
Mixed	2.79	1.018	2.65	0.966	2.54	0.929	2.23	0.814	2.47	0.903
Reference	3.51	1.283	3.34	1.221	3.10	1.132	2.96	1.083	3.14	1.145

The 3-year period is the mean of the 3 study years. This 3-year period was synthesized because the study-period average combined 3 growing seasons with 2 non-growing seasons and, therefore, is biased high compared to an annual average. Little uncertainty is introduced by the substitutions because ET is small during the non-growing season. As seen in table 11, study period ET was about 5 percent greater at the bulrush site than at the mixed site. Because the EC data were corrected to close the energy balance, the two factors affecting this 5 percent difference are mean values of available energy, AE, and how that energy was partitioned between H and LE, or the Bowen ratio, β. Mean values of AE during the study period were 127.0 and 115.1 W m^{-2} at the bulrush and mixed sites respectively. (These values differ slightly from the R_n means cited in the previous paragraph by the amounts of mean $Q_x + G$, which were -0.6 and -1.2 W m^2 at the bulrush and mixed sites, respectively.) Although AE is about 10 percent greater at the bulrush site than at the mixed site, a greater proportion of AE was partitioned into LE at the mixed site, reducing the LE (and ET) site difference to about 5 percent. As noted earlier, the greater proportion of open water associated with stands of wocus at the mixed site possibly resulted in a lower mean β at the mixed site (0.450) than at the bulrush site (0.525) despite the 0.15-m higher land surface and resulting shallower water depths at the mixed site.

A two-sided paired-t test was conducted to determine whether daily ET values from the two sites during the study period are statistically different at the $\alpha = 0.05$ probability level. Site ET differences are normally distributed. This test yielded a t-value of 2.067 ($p = 0.0369$), indicating that the null hypothesis can be rejected and ET values from the two sites are statistically different, but not conclusively. The accuracy of EC data is typically given as ±5 to 10 percent under ideal conditions (Foken, 2008) such as found here, and the corresponding precision (or ability to distinguish differences under similar conditions) should be substantially smaller than the accuracy, on the order of ±1 to 2 percent. Therefore it seems likely, although not conclusively, that ET was slightly and significantly greater from the bulrush site than from the mixed site.

Variability between years (about ± 8.5 percent) was substantially greater than variability between sites (about ± 2.5 percent). The years of 2008 and 2009 were very similar overall; bulrush ET was slightly greater in 2009 than in 2008, whereas mixed site ET was slightly greater in 2008 than in 2009 (table 11). In 2010, ET at both sites was about 14 to 15 percent less than during the previous 2 years. Available energy at the bulrush site was 1 percent greater in 2010 than in 2008 through 2009. Available energy at the mixed site was 2 percent less than in 2008 through 2009, and therefore it cannot account for the lower ET in 2010. As

discussed earlier, unusually low water levels in the first half of 2010 probably reduced evaporation from the standing water by decoupling the water surface from incoming radiation. In addition, synthesized precipitation (P) was 16 percent lower in 2010 than in 2008–2009 and may have reduced evaporation of intercepted water, especially during the non-growing season. (Although the last 3 months of the synthesized 2010 data are borrowed from 2009, the substitution was made for ET as well as P and should preserve any correlation between ET and P.) Therefore, lower water levels and precipitation probably were the main factors causing the lower ET in 2010.

Maximum annual ET occurred in 2009 at the bulrush site and in 2008 at the mixed site. Differences between the maximum annual value and the next lower value are small (1.8 and 4.0 percent at the bulrush and mixed sites, respectively), and the year reversal could be caused either by site differences in climate and surface response to climate, or simply by random error in the EC method. However, the most likely causative climatic variable, available energy, was greater at both sites in 2009, suggesting the year reversal may be a result of random error.

Overall, the 3-year synthesized average ET values at the bulrush and mixed sites are 0.953 m/yr and 0.903 m/yr, respectively. The corresponding 3-year synthesized average ET_r from the Agency Lake weather station is 1.145 m/yr, resulting in average annual K_c values of 0.832 and 0.789 at the bulrush and mixed sites, respectively. These annual values fall near the middle of the range of ensemble average biweekly K_c values given in figure 17 and table 8. Based on our estimate that 70 percent of the Upper Klamath NWR is typified by the bulrush site and 30 percent by the mixed site, the 3-year average ET value for the whole NWR is 0.938 m/yr. The 3-year synthesized average precipitation at the Klamath Falls weather station (KFLO) is 0.233 m/yr, equal to about 25 percent of 3-year average measured ET.

The relation of the 3-year synthesized ET values to expected values of long-term ET can be partially investigated by comparing relevant climatic variables during the 3-year period to their long-term means. In a water-limited system, precipitation (P) is probably the most important variable, whereas in an energy-limited system, solar radiation (R_s) and temperature assume that role. Although the most relevant temperature is that of the evaporating surface, air temperature (T_a) is commonly used because it is highly correlated with surface temperature and is more readily available. During the heart of the growing season, the NWR wetland clearly is energy-limited, but at other times of the year, P influences ET through evaporation of intercepted water. Long-term means of R_s, P, and T_a are calculated using data from the nearby Agency Lake (AGKO) and Klamath Falls (KFLO) weather stations (fig. 1) and are compared to 3-year synthesized study period means in table 12. Long-term means are calculated from water years 2001–2011 (11 years) and the 3-year values are synthesized from the same months and years used to compute measured ET, to be consistent with those data. The closer AGKO station is used when possible (for R_s and T_a), but only the KFLO station has a reliable precipitation record during the winter, so those data are used to compute P. Although the KFLO station record begins in 1999, the first 2 years are not used, to be consistent with the full period of record at the AGKO station (2001–2011). Solar radiation during the study period was about 9.5 percent lower than the 11-year mean, which should reduce the study period ET compared to the long term. Interestingly, P also was about 17 percent less during the study period than its long-term average. Often, reduced P is associated with greater R_s, but the frequency, duration, and intensity of P, as well as the distance between the AGKO and KFLO stations, can alter this relation. In this wetland setting, reduced P should have little effect on growing-season ET, and a mildly inhibiting effect on non-growing season evaporation of intercepted water. Finally, the study-period T_a is slightly cooler than the long-term mean, which also should slightly reduce ET. Overall, the study period was less sunny, drier, and cooler than the 11-year average, all of which should reduce the study-period ET compared to the norm. While this result indicates that the study-period ET given in table 11 probably is smaller than the long-term average, the smaller ET should have little or no effect on the accuracy of the K_c approach developed in this study. The effects of R_s and T_a on ET are explicitly included in the Penman-Monteith expression for ET_r, and the effects of P on annual ET from this well-watered wetland are small.

Table 12.   Comparison of 3-year study period averages with 11-year averages of solar radiation, precipitation, and air temperature.

[The 3-year period is synthesized from the actual May 1, 2008–September 29, 2010 study period, supplemented with data from January 1, 2009–April 30, 2009 and September 30, 2009–December 31, 2009, to be consistent with synthesized 3-yr evapotranspiration (ET) values given in table 11. Solar radiation and air temperature data are from the Agency Lake (AGKO) weather station, and precipitation data are from the Klamath Falls (KFLO) weather station. **Abbreviations**: mm yr^{-1}, millimeters per year; W m^{-2}, watts per square meter; °C, degrees Celsius]

Period	Solar radiation (W m^{-2})	Precipitation (mm yr^{-1})	Air temperature (°C)
2008–2010	184.7	233	6.98
2001–2011	195.0	280	7.45

Comparison of Wetland Evapotranspiration with Evapotranspiration of Local Crops

A comparison of our measured wetland ET with ET of crops in the area can be made using data from the Bureau of Reclamation AgriMet Web site. Computation of ET_r is based on an assumption that the alfalfa crop is full sized and green year-round, as indicated by an assumed surface resistance (r_s) of 45 s/m at the daily time step, regardless of time of year (Allen and others, 2005). In the region of the current study, actual alfalfa ET is somewhat less than ET_r due to reduced green leaf area during the non-growing season and multiple cuttings during the growing season. The Bureau of Reclamation AgriMet Web site computes and posts values of estimated ET for alfalfa and other crops. These ET values are computed using the 1982 Kimberly-Penman equation to compute ET_r, with appropriate values of K_c. Although this ET_r equation is slightly different than the ASCE equation, the K_c values were developed to provide best estimates of crop ET using the 1982 ET_r equation (the AgriMet Web site does not provide estimates of crop ET using the more recent ASCE ET_r equation). The Web site posts crop ET for each growing season, the duration of which is established by local conditions and expert opinions year by year. We computed wetland ET using our measured data during the 2008 through 2010 growing seasons, and the 70 percent to 30 percent weighting (for bulrush versus mixed vegetation) used earlier in Study-Period and Annual Evapotranspiration. We obtained alfalfa and pasture growing-season ET data for 2008 through 2010 from the Klamath Falls (KFLO) AgriMet Web site because crop ET is not computed for the closer AGKO station. A comparison of growing-season ET (table 13) shows that wetland ET was about 9 percent less than alfalfa ET, and about 18 percent greater than pasture ET during 2008 through 2010.

Although AgriMet does not compute non-growing-season crop ET, a rough comparison of annual wetland and crop ET can be made by assuming ET from all vegetated surfaces is equal during the non-growing season. This assumption is supported by the similar weather and existence of dormant vegetation at the locations involved. Although wetland ET probably exceeds crop ET in late winter and early spring due to the presence of standing water at the wetland, crop ET probably exceeds wetland ET during fall and early winter (when standing water is absent) because of the greater mulching effect of the more massive dormant canopy at the wetland. We compute mean 2008 through 2010 non-growing season wetland ET using the same synthesis methods used in Study-Period and Annual Evapotranspiration to estimate early 2008 and late 2010 water use. Non-growing season crop ET is assumed equal to wetland ET, and annual ET is computed as the sum of growing-season and non-growing-season ET of both wetland and crops. On an annual basis, wetland ET during 2008 through 2010 is estimated to be about 6 percent less than alfalfa ET and about 14 percent greater than pasture ET (table 13).

Table 13.   Comparison of mean 2008–2010 growing season, non-growing season, and annual evapotranspiration (*ET*) of alfalfa, pasture, and Upper Klamath Lake NWR wetland, in meters.

[Alfalfa and pasture growing-season duration and *ET* determined from Bureau of Reclamation AgriMet Web site. Wetland *ET* measured using eddy covariance method (this study). Non-growing season alfalfa and pasture *ET* assumed equal to wetland *ET* during the same period. Annual *ET* is the sum of growing season and non-growing season *ET*. **Abbreviation**: NWR, National Wildlife Refuge]

	Alfalfa comparison (190 day average growing season)		Pasture comparison (195 day average growing season)	
	Alfalfa *ET*	Wetland *ET*	Pasture *ET*	Wetland *ET*
Growing season	0.838	0.779	0.671	0.789
Non-growing season	0.159	0.159	0.149	0.149
Annual	0.997	0.938	0.820	0.938

Comparison of Wetland Evapotranspiration with Previous Studies

Bidlake (2000) conducted a field campaign to characterize *ET* at a site about 200 m northwest of the bulrush site during the growing season of 1997. He deployed EC sensors during four 1- to 2- day periods (May 29–30, July 10–12, August 23–25, and October 11–13), and a recording weather station at the same location continuously from May 29 to October 13. The EC fluxes were used to calibrate a Penman-Monteith (PM) model (similar to the one used in the present study to compute ET_r at the AGKO AgriMet site) by solving for the average surface resistance, r_s, during each EC deployment, and interpolating r_s between site visits. The Penman-Monteith model was then used to compute daily *ET* during the entire growing season, and the daily values were aggregated into weekly values. Bidlake's EC data were not corrected to force energy-balance closure and, therefore, neither were the modeled *ET* values because they were calibrated to match the EC data. For comparison with the current study, we divided Bidlake's *ET* data by the average energy budget ratio (*EBR*) measured during that study (0.88) to force energy-balance closure and improve comparability with the current study.

Bidlake's 1997 adjusted weekly *ET* (Bidlake, 2000) is plotted along with our 2008 through 2010 biweekly bulrush *ET* in figure 19. Compared to our 2008 and 2009 ET data, the 1997 *ET* was somewhat lower in June, nearly equal in July, somewhat greater in August, and nearly equal thereafter, resulting in very similar growing season totals. June and August differences may be attributable to: interpolation of r_s to compute the 1997 *ET*; the use of relatively short periods (29–35 hours) to compute each r_s interpolation endpoint; or to climatic differences between years. Worthy of note is that in 1997, June appeared to be cloudier, and August

clearer, than the corresponding months during the current study (fig. 10; Bidlake, 2000, fig. 1), which is consistent with the *ET* differences during those periods. Interestingly, *ET* magnitudes during all 4 years are similar from late August to early October, a time when the canopy is tall and beginning to senesce. Overall, the 1997 growing season *ET* data are in general agreement with the *ET* measured in the current study and help support our results.

During the growing season of 2000, Bidlake (2002) measured *ET* from three fallowed agricultural fields in the Tule Lake National Wildlife Refuge in northern California, about 83 km southwest of the present study site. During the previous year (1999), the three sites were sown to barley and irrigated, and had variously evolved into a patchwork of harvested, unharvested, and burned (by wildfire) areas by spring 2000. One of the sites was tilled and sown to rye in February 2000 to simulate a dry-land cover crop that might be used during future fallowing. During the growing season of 2000, surface coverage consisted of a mixture of stubble and other crop residues, weedy volunteer species, small grain plants sprouted from the seeds of previous crops, bare soil, and rye (at the one site planted with rye). All sites were un-irrigated during 2000. Water table depths averaged about 1.1 m below land surface during the growing season, ranging from about 0.7 to 1.4 m. Evapotranspiration was measured or estimated from May 1 to October 31 (184 days, designated as the growing season) using the Bowen-ratio method, supplemented with modeling during data gaps. The Priestley-Taylor method (with variable α) was used during short (< 2 hr) gaps, and a reference *ET* (ET_r) method was used for longer gaps. Daily ET_r was obtained from a University of California weather station about 8 km north, and daily K_c values were computed by interpolation between, or extrapolation beyond periods when Bowen-ratio sensors were operational. Extended gaps occurred at the beginning and end of the growing season, approximately the

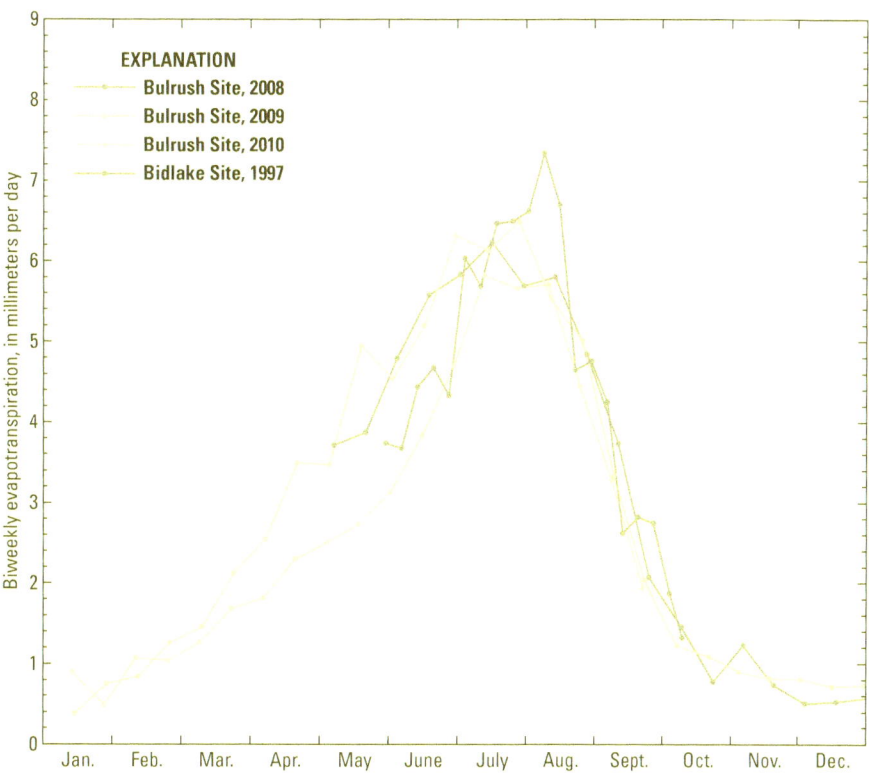

Figure 19. Comparison of biweekly evapotranspiration (ET) measured at the bulrush site during the current study with weekly ET measured and modeled by Bidlake (2000) at a nearby site during May 28–October 12, 1997.

first half of May, and the last two-thirds of October, at all 3 sites. Because ET at these times was small, little error in the seasonal totals was incurred by the use of the ET_r model. Although timing of peak ET varied substantially between the fields, the growing-season totals were quite similar, ranging from 0.426 to 0.444 m, and averaging 0.435 m.

We computed 2008 through 2010 wetland growing-season ET using our measured data from the same growing season as in Bidlake (2002), again weighting the bulrush and mixed site ET 70 percent-30 percent, respectively. Growing-season ET data from October 2009 were substituted for the October 2010 period, which occurred after the study period. Wetland growing-season ET was about 65 percent greater than ET from the fallowed cropland (table 14). Although Bidlake did not compute non-growing season ET at the fallowed crop sites, a rough comparison of annual wetland and fallowed-crop ET can be made by assuming ET from both vegetated surfaces is equal during the non-growing season, as was assumed for alfalfa and pasture. We compute mean 2008 through 2010 non-growing season wetland ET using the same synthesis methods used in Study Period and Annual Evapotranspiration to estimate early 2008 and late 2010 water use. Non-growing season fallowed-crop ET is assumed equal

to wetland ET, and annual ET is computed as the sum of growing-season and non-growing-season ET of both wetland and fallowed crops. On an annual basis, wetland ET during 2008–2010 is estimated to be about 43 percent greater than fallowed-cropland ET during 2000 (table 14).

Table 14. Mean growing season, non-growing season, and annual evapotranspiration (ET) of Tule Lake NWR fallowed cropland during 2000, and Upper Klamath Lake NWR wetland during 2008–2010, in meters.

[Fallowed cropland growing-season ET is from Bidlake (2002). Wetland ET measured using eddy covariance method (this study). Non-growing season fallowed cropland ET assumed equal to wetland ET during the same time of year. Annual ET is the sum of growing season and non-growing season ET. **Abbreviations:** NWR, National Wildlife Refuge]

	Tule Lake NWR fallowed crops	Upper Klamath Lake NWR wetland
Growing season	0.435	0.718
Non-growing season	0.220	0.220
Annual	0.655	0.938

Open-Water Evaporation Results

Average open-water evaporation rates during the energy budget periods ranged from 2.8 mm d^{-1} in early October 2008 at the MDL site (MDN was not usable in 2008) to as much as 7.0 mm d^{-1} during mid-July 2010 at the MDL site (table 15). Evaporation totals from June 12, 2009, to October 2, 2009, were 615 and 601 mm at the MDL and MDN sites, respectively. The difference between totals at the sites was about 2.3 percent. Evaporation during the same period in 2010 was slightly larger, totaling 636 and 611 mm at MDL and MDN, respectively. Evaporation totals for the entire period

of data for 2010, May 29 through October 2, which was two weeks longer than the data collection period in 2009, were 701 and 675 mm at the MDL and MDN sites, respectively. The difference between the 2010 totals at the two sites was about 3.8 percent.

Open-water evaporation measurements generally correspond to the period from June through September. About 61 percent of annual pan evaporation at Klamath Falls occurs during this period, based on mean monthly values from 1949 to 2004 (Western Regional Climate Center, 2012). The surface of Upper Klamath Lake is typically frozen during winter months.

Table 15.   Mean surface temperatures (T_s), Bowen ratios (β), net radiation (R_n), net water advected energy (Q_v), energy transferred to lakebed (Q_b), net change in energy stored in the lake (Q_x), and energy-budget evaporation rates (E_{eb}) for biweekly budget periods in 2008, 2009, and 2010.

[**Abbreviation**: MDL, midlake site; MDN, midlake north site; mm d^{-1}, millimeters per day; W m^{-2}, watts per square meter;   C, degrees Celsius]

Year	Budget period dates	Budget period number	T_s (MDL) (°C)	T_s (MDN) (°C)	β (MDL)	β (MDN)	R_n (W m^{-2})	Q_v (W m^{-2})	Q_b (W m^{-2})	Q_x (W m^{-2})	E_{eb} (MDL) (mm d^{-1})	E_{eb} (MDN) (mm d^{-1})
2008	07-24–08-06	7	21.8		0.102		211.73	-8.64	4.07	1.21	6.2	
	08-07–08-20	8	22.0		0.072		174.35	-9.86	2.57	-19.27	5.8	
	08-21–09-03	9	18.9		0.159		158.64	-9.30	0.92	-31.28	5.3	
	09-04–09-17	10	18.2		0.024		132.33	-6.49	-0.79	2.87	4.1	
	09-18–10-01	11	16.2		0.101		92.47	-4.31	-2.44	-18.00	3.4	
	10-02–10-15	12	11.4		0.296		48.74	-1.36	-3.96	-51.83	2.8	
2009	06-12–06-25	17	18.7	18.5	0.214	0.221	220.43	-2.11	6.78	8.99	5.7	5.7
	06-02 –07-09	18	21.0	21.5	0.098	0.120	223.22	-11.52	6.17	2.41	6.3	6.2
	07-10–07-23	19	22.0	22.2	0.070	0.092	229.65	-11.42	5.20	19.00	6.2	6.1
	07-24–08-06	20	23.6	24.5	0.060	0.098	179.06	-13.01	3.94	-37.24	6.4	6.2
	08-07–08-20	21	19.8	20.3	0.110	0.139	192.02	-8.99	2.44	-0.09	5.6	5.4
	08-21–09-03	22	20.4	20.5	0.065	0.100	158.84	-11.65	0.81	-17.46	5.3	5.1
	09-04–09-17	23	18.2	18.9	0.108	0.155	135.12	-7.19	-0.87	-16.14	4.5	4.3
	09-18–10-01	24	17.1	17.5	0.144	0.171	98.92	-6.47	-2.50	-34.35	3.9	3.8
2010	05-29–06-11	29	14.0	13.9	0.187	0.193	189.62	4.32	6.82	27.82	4.6	4.6
	06-12–06-25	30	17.9	16.8	0.230	0.192	233.84	1.48	6.82	31.70	5.5	5.7
	06-26–07-09	31	19.2	20.0	0.111	0.138	234.69	-3.70	6.43	12.86	6.5	6.4
	07-10–07-23	32	21.8	22.6	0.046	0.075	231.08	-5.92	5.66	5.24	7.0	6.8
	07-24–08-06	33	22.9	23.1	0.084	0.109	197.11	-4.79	4.57	-9.81	6.2	6.1
	08-07–08-20	34	22.4	22.3	0.096	0.111	188.66	-7.61	3.21	-13.62	6.0	5.9
	08-21–09-03	35	17.0	19.2	0.061	0.181	158.97	-6.47	1.67	-26.14	5.7	5.2
	09-04–09-17	36	15.0	17.1	0.025	0.194	141.93	-4.01	0.03	-14.46	5.1	4.4
	09-18–10-01	37	16.6	16.6	0.165	0.219	112.48	-0.95	-1.61	0.58	3.3	3.2

During the study period, water surface temperatures of the lake were highest from late July through the first week of August. In 2008, however, the highest biweekly average surface temperature occurred during the period that extends through mid-August (table 15). Peaks in average surface temperatures tended to lag peaks in net radiation (fig. 20; table 15). Energy-budget evaporation rates are generally at their highest when net radiation is highest (fig. 21), but rates are also affected by other factors, chiefly net changes in energy stored in the lake and the dominant type of heat transfer occurring in the lake as described by Bowen ratios (table 15).

Where data were available to calculate evaporation at both the MDL and MDN sites, differences in evaporation rates averaged about 0.2 mm/d, and the median difference for all energy budget periods was 0.1 mm/d. The maximum difference observed was 0.7 mm/d (table 15). The similarity of evaporation rates determined at the geographically separate MDL and MDN sites indicates that, at the time scales of the 2-week energy-budget periods, the physical conditions controlling the Bowen ratio were similar at both sites. Given that water depth and fetch conditions at these two sites are representative of most of the lake, and that the lake is reasonably well mixed, results from the MDL and MDN sites should be reasonably representative of the lake as a whole.

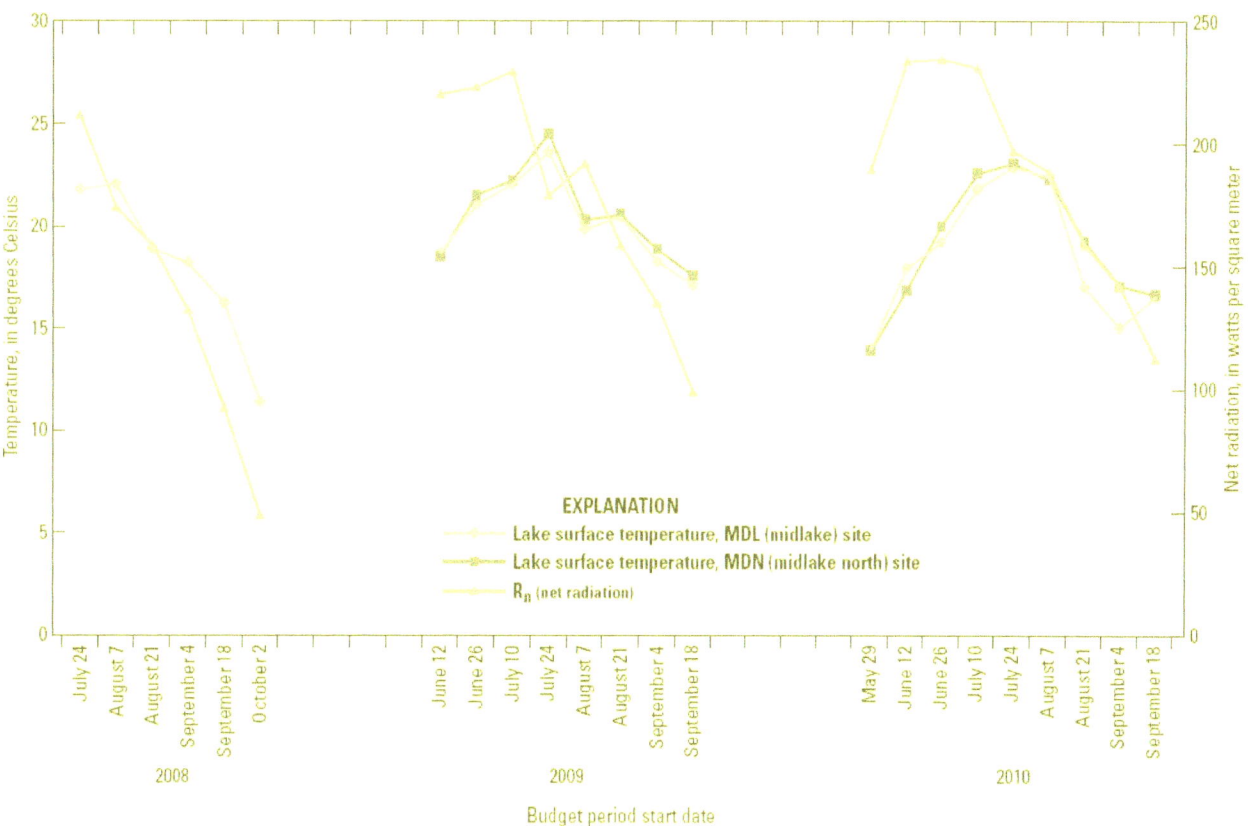

Figure 20. Average lake surface temperatures at the midlake (MDL) and midlake north (MDN) sites plotted on the left vertical axis and net radiation (Q_n) plotted on the right vertical axis for the budget periods extending from July 2008 to September 2010. Changes in surface temperatures generally lag changes in net radiation.

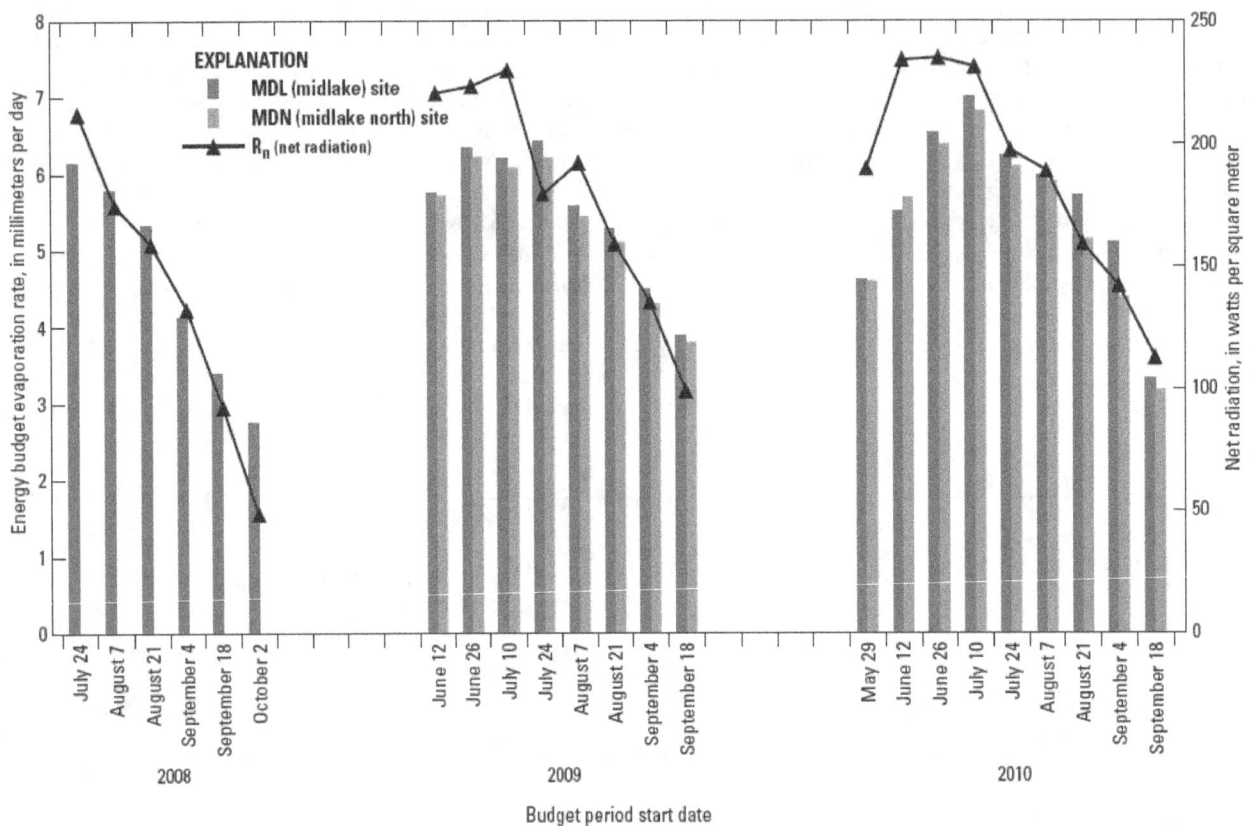

Figure 21.   Average evaporation rates at the midlake (MDL) and midlake north (MDN) sites plotted on the left vertical axis and net radiation (Q_n) plotted on the right vertical axis for the budget periods from July 2008 to September 2010.

Comparison of Open-Water Evaporation with Previous Studies and With Wetland Evapotranspiration

Biweekly energy-budget evaporation rates calculated for budget periods in 2008, 2009, and 2010 are similar to evaporation rates determined in previous studies of Upper Klamath Lake. Janssen (2005) determined that 2003 energy budget evaporation rates from June 7 to September 30 averaged 4.2 mm/d, somewhat less than the 5.5 mm/d average measured during a similar period in 2009 and 2010 in this study in spite of slightly warmer monthly average temperatures in 2003. Janssen's average, however, excludes data from 2 days in late June along with a period of 13 days between June 26 and July 8. Because near-peak evaporation occurs during these missing periods, Janssen's average probably would have been greater if the missing data had been available.

Using a 1-D surface energy balance model to produce simulations of Upper Klamath Lake daily evaporation for 1950 through 2005, Hostetler (2009) found average May to September evaporation totaled 707 mm. This compares favorably with the late-May to September 2010 totals of 698 and 680 mm measured at the MDL and MDN sites, respectively, during this study.

Seasonal trends in open-water evaporation and wetland *ET* were similar (figs. 15, 21). In general, open-water evaporation exceeded wetland *ET* during the periods when both were measured. As expected, *E–ET* was greatest during the late summer periods (late August to October), due to the greater release of stored heat in the lake compared to the land surface. Differences were smaller during midsummer (late June to early August), a time when vegetation was at full height and stored heat was not yet a factor. A notable exception occurred in late June through mid-July 2010, when *E–ET* was unusually large. The low lake levels in 2010 probably contributed to this difference through reduced evaporation of standing water from the wetland, as discussed in Daily Evapotranspiration and Crop Coefficients. Early-season (late May to mid-June) data are spotty (only three periods), but consistently show substantially greater open-water evaporation, as the green vegetation canopy has not fully emerged from the dead stalk mat, and replaced it as the primary exchange surface. Overall, during the periods of open-water data collection, open-water evaporation (the mean of the MDL and MDN sites) was 20 percent greater than wetland *ET* (a 70 percent-30 percent weighted mean of the bulrush and mixed sites, respectively).

Summary and Conclusions

Water allocation in the Klamath Basin is important to many different groups of people, and to the health and well-being of many ecosystems within the basin. Competition for the limited water supply in this semiarid basin is intense, and knowledge of water losses to evapotranspiration (*ET*) is key information for deciding optimal water-use strategies. One of the central hubs for water distribution in the basin is Upper Klamath Lake, near Klamath Falls, Oregon. The U.S. Geological Survey (USGS) and the Bureau of Reclamation conducted a study to quantify *ET* from extensive wetlands in the Upper Klamath National Wildlife Refuge (NWR) at the northwest periphery of the lake, and evaporation from the open-water portion of the lake. Data collection spanned the period from May 1, 2008, to September 29, 2010.

Two wetland sites were selected for study to typify vegetation communities and hydrologic conditions most frequently occurring in the NWR. Vegetation at one wetland site consisted of a virtual monoculture of bulrush (formerly *Scirpus acutus*, now *Schoenoplectus acutus*), while the other site was a mixture dominated by bulrush, with minor amounts of cattail (*Typha latifolia*), wocus (*Nuphar polysepalum*), open water, and trace amounts of other vegetation. The bulrush site is at an altitude of 1,262.03 m (4,140.02 ft) asl and the mixed site is about 0.15 m (0.50 ft) higher. As a result of controlled lake levels, the wetlands are periodically inundated with lake water on an annual basis. Standing water typically occurs from January or February through July or August, for an average hydroperiod of about 6 months at the bulrush site and 5 months at the mixed site. Water level typically fluctuates about 1.3 m, from about 0.8 to 0.9 m above land surface at the bulrush site in spring and early summer to about 0.4 to 0.5 m below land surface in October, with correspondingly lower levels at the mixed site. Minimum lake level in fall 2009 was unusually low, resulting in a late water-level rise and reduced maximum water level in 2010. The root zone of the bulrush site remained saturated during the entire study period (volumetric water content, θ, was 0.8–0.9 m³/m³), whereas the mixed site root zone partially dewatered to a θ of around 0.5 m³/m³ during late summer and re-saturated in late winter or spring. Canopy height at both sites typically reaches a maximum of 2.2 to 2.3 m during the summer, although maximum height at the mixed site in 2010 was only 1.9 m. When the canopy senesces in October, the dead stalks remain vertical for some time, and eventually lodge over in response to snow and wind loading. The bent-over stalks form a loosely distributed mat, generally about 0.6 to 1.0 m high.

The eddy-covariance (EC) method was used to measure *ET* from two wetland sites. A source-area model estimates that at the bulrush site, 98 to 99 percent of the measured *ET* originated from within the wetland, and at the mixed site, 95 to 96 percent originated from within the wetland, indicating that contamination of the EC measurement by other surface types was insignificant. A chronic inability to close the surface energy balance using the EC method was remedied by adjusting the turbulent flux upward to equal the long-term available energy, while maintaining the ratio between the sensible- and latent-heat fluxes (the Bowen ratio, or β). This common adjustment (Foken, 2008) made the estimates of wetland and open-water evaporation consistent with one another.

Partitioning of available energy (*AE*) into sensible-heat flux (*H*) and latent-heat flux (*LE*) varies dramatically during the seasonal cycle (*LE* is the energy equivalent of the *ET* rate). At the beginning of the calendar year, the remnant dead vegetation mat from previous years is the dominant exchange surface with the atmosphere and solar radiation, although rising water levels soon overtop the land surface, forming a secondary, mostly shaded water surface beneath the dead canopy. When the canopy is dry, β tends to be greater than one. Precipitation briefly shifts the balance toward *LE* for one to a few days, dropping β below 1, but evaporation and infiltration of intercepted water to lower layers soon return the partitioning to $\beta > 1$. As a result of these repeated rapid reversals, on average, $\beta \approx 1$ during most of the winter. As *AE* increases with growing sun angles in the spring, both *H* and *LE* grow in response, maintaining a β of around one. The new green transpiring vegetation typically emerges from the water and dead stalk mat in May or June, at which time partitioning begins to shift rapidly toward *LE*. By July *LE* has reached peak values, and *H* has actually decreased from its spring values, even though *AE* has increased to peak values. Through much of the growing season, *LE* is much greater than *H*, as indicated by β values around 0.26 at the bulrush site and 0.13 at the mixed site. Possibly the greater partitioning toward *LE* at the mixed site is in response to the greater proportion of open water at that site, in spite of lower standing water levels. As the vegetation canopy senesces during September, transpiration decreases, and partitioning ceases to favor *LE*. During the study, the Octobers were unusually dry, receiving 52 percent of mean 1999 through 2011 precipitation at Klamath Falls. This lack of precipitation and transpiration, coupled with water levels below land surface, typically resulted in β of around 2 or 3. Greater precipitation in November and December (the two greatest precipitation months historically) again restores the average β to around one.

Measured *ET* at the wetland sites is compared to reference *ET* (ET_r) computed from data collected at the nearby Bureau of Reclamation Agency Lake weather station (AGKO), to compute crop coefficients (K_c) at daily, biweekly, and annual time steps. Approximate formulas are given to estimate daily values of growing season K_c, thereby allowing computation of daily *ET* using ET_r from the AGKO weather station. Biweekly values of growing season K_c are computed from ensemble average values of *ET* and ET_r during the 3-year study period growing seasons, and a single, mean value of K_c is computed for the non-growing season. Together, these provide relatively accurate estimates of biweekly *ET* during the study (*RMSE*=0.396 and 0.347 mm d^{-1}, r^2=0.962 and 0.971 at the bulrush and mixed sites, respectively). A fourth-order polynomial fit of the growing season values to day of year provides a more automated form of *ET* computation.

Annual values of *ET* from both study sites are computed for the calendar years of 2008, 2009, and 2010. Annual values for 2008 and 2010 are synthesized by substituting non-growing season *ET* data from 2009 for periods before the study began (January 1–April 30, 2008) and after the study ended (September 30–December 31, 2010). Little uncertainty is introduced by this substitution because *ET* is small during these times. Annual *ET* at the bulrush site ranged from 0.850 to 1.013 m/yr, with a 3-year mean of 0.953 m/yr (table 11). Annual *ET* at the mixed site ranged from 0.814 to 0.966 m/yr, with a 3-year mean of 0.903 m/yr. Based on a satellite-image estimate that the bulrush site typifies 70 percent of the Upper Klamath Lake National Wildlife Refuge (NWR) and the mixed site typifies 30 percent, the resulting estimate of mean NWR *ET* during the 3 years is 0.938 m/yr. At both sites, minimum annual *ET* occurred during 2010, a year of unusually low water level during spring and early summer, and a year that had 16 percent less precipitation than in 2008 through 2009. Although the lower 2010 water level almost surely did not stress the vegetation, the water surface remained more shaded and decoupled from the atmosphere, leading to less surface water evaporation. Interannual variability (about ± 8.5 percent) was substantially greater than intersite variability (about ± 2.5 percent). A paired t-test conducted on site differences indicated that daily *ET* values from the two sites were statistically, but only marginally different from one another at the α=0.05 level.

Study-period averages of solar radiation (R_s), air temperature (T_a), and precipitation (*P*) measured at nearby AgriMet stations all were lower than their corresponding 11-year averages from 2001–2011. Normally, lower R_s is associated with greater *P*, but in this case, the study period was less sunny and drier than the long-term average. These climatic conditions all suggest that *ET* measured during the study period probably is less than the expected long-term average. However, because R_s and T_a appear in the expression

for reference *ET* (ET_r), and the effect of *P* on *ET* in this wetland setting is minor, the crop coefficient approach developed here should remain relatively robust in other climatic settings.

A comparison of measured growing-season wetland *ET* with alfalfa and pasture *ET* was made by tabulating alfalfa and pasture *ET* totals posted for the Klamath Falls station (KFLO) on the Bureau of Reclamation AgriMet Web site (http://www.usbr.gov/pn/agrimet/ETtotals html) and obtaining growing season durations (http://www.usbr.gov/pn/agrimet/chart/kflo08et.txt). During the 190-day average alfalfa growing seasons from 2008 through 2010, wetland *ET* at our study sites was 0.779 m, about 7 percent less than alfalfa mean *ET* of 0.838 m. During the 195-day average pasture growing seasons from 2008 through 2010, wetland *ET* at our study sites was 0.789 m, about 18 percent greater than pasture mean *ET* of 0.671 m. A comparison of annual *ET* can be made by assuming alfalfa and pasture *ET* are equal to wetland *ET* during the non-growing season. This assumption leads to annual estimates for 2008 through 2010 of 0.997, 0.938, and 0.820 m of *ET* from alfalfa, wetland, and pasture land covers, respectively.

An earlier *ET* study of this wetland used short-term (1- to 2-day) EC measurements of *ET* made four times during the growing season of 1997 to calibrate a Penman-Monteith *ET* model, driven by continuous onsite weather data, to compute growing-season *ET* (Bidlake, 2000). The study site was located about 200 m northwest of our bulrush site, with virtually identical vegetation and altitude. Total growing-season *ET* from that study was comparable to our 2008 and 2009 *ET*, with notable short-term differences that appear to be related to differences in cloudiness. The two studies are consistent, and together they provide a well-documented estimate of Upper Klamath NWR evapotranspiration.

Bidlake (2002) also studied *ET* from fallowed cropland in the Tule Lake NWR during the growing season of 2000, using the Bowen-ratio energy balance method supplemented with modeling to fill in data gaps when sensors malfunctioned, and to extend the period of record to the full growing season (defined as May 1–October 31 in that study). Although the seasonal timing of *ET* varied considerably among the three types of fallow surfaces studied, total *ET* was remarkably constant, at 0.435 ± 0.009 m during the growing season. During the same months, mean *ET* measured at the Upper Klamath Lake NWR wetland in the current study was 0.718 m, or about 65 percent greater than the fallowed-cropland *ET*. If the non-growing-season *ET* measured at the wetland (0.220 m) also is assumed to occur from the fallowed cropland, the resulting annual *ET* from the wetland during 2008 through 2010 (0.938 m) is about 43 percent greater than the fallowed cropland *ET* (0.655 m) during 2000.

Open-water evaporation was measured at two midlake locations during 23 warm season biweekly periods during the study, using the Bowen-ratio energy balance method. Seasonal patterns of open-water evaporation were similar to those of wetland *ET*, but exhibited less of a seasonal cycle. Open-water and wetland magnitudes were nearly equal during late June to early August, when wetland vegetation was green and abundant. During late summer, open-water evaporation consistently exceeded wetland *ET* as the vegetation canopy began to senesce, while the lake released much of the energy stored during the summer, as latent-heat flux. Spring comparison data are few, but suggest that open-water evaporation exceeds wetland *ET* when emergence of the new vegetation canopy out of the old, dead stalk mat is not complete. Overall, measured open-water evaporation was 20 percent greater than wetland *ET* during the same periods.

References Cited

Allen, R.G., Pereira, L.S., Raes, D., and Smith, M., 1998, Crop evapotranspiration—Guidelines for computing crop water requirements: United Nations Food and Agriculture Organization Irrigation and Drainage Paper no. 56, 300 p.

Allen, R.G., Walter, I.A., Elliott, R.L., Howell, T.A., Itenfisu, D., and Jensen, M.E., 2005, The ASCE standardized reference evapotranspiration equation: Reston, Va., American Society of Civil Engineers, 196 p.

Anderson, E.R., 1954, Energy-budget studies, *in* Harbeck, G.E., and others, Water-loss investigations: Lake Hefner studies: U.S. Geological Survey Professional Paper 269, p. 71–119.

Aubinet, M., Vesala, T., and Papale, D., 2012, Eddy covariance—A practical guide to measurement and data analysis: Dordrecht, Germany, Springer, 438 p.

Barr, A.G., Betts, A.K., Desjardins, R.L., and MacPherson, J.I, 1998, Comparison of regional surface fluxes from boundary layer budgets and aircraft measurements above boreal forest: Journal of Geophysical Research, v. 102, no. D24, p. 29213–29218.

Bidlake, W.R., 2000, Evapotranspiration from a bulrush-dominated wetland in the Klamath Basin, Oregon: Journal of the American Water Resources Association, v. 36, no. 6, p. 1309–1320.

Bidlake, W.R., 2002, Evapotranspiration from selected fallowed agricultural fields on the Tule Lake National Wildlife Refuge, California, during May to October 2000: U.S. Geological Survey Water-Resources Investigations Report 02–4055, 59 p.

Blanken, P.D., Black, T.A., Yang, P.C., Neumann, H.H., Nesic, Z., Staebler, R., den Hartog, G, Novak, M.D., and Lee, X., 1998, Energy balance and canopy conductance of a boreal aspen forest: Partitioning overstory and understory components: Journal of Geophysical Research, v. 102, no. D24, p. 28915–28927.

Bossong, C.R., Caine, J.S., Stannard, D.I., Flynn, J.L., Stevens, M.R., and Heiny-Dash, J.S., 2003, Hydrologic conditions and assessment of water resources in the Turkey Creek Watershed, Jefferson County, Colorado, 1998–2001: U.S. Geological Survey Water-Resources Investigations Report 03–4034, 140 p.

Bowen, I.S., 1926, The ratio of heat losses by conduction and by evaporation from any water surface: Physical Review, v. 27, p. 779–787.

Brutsaert, W., 1982, Evaporation into the atmosphere: Boston, D. Reidel Publishing Co., 299 p.

Bureau of Reclamation, 2005, Natural flow of the upper Klamath River—Phase I: Prepared by the Technical Service Center, Denver, Colo., for U.S. Department of the Interior, Bureau of Reclamation, Klamath Basin Area Office, Klamath Falls, Oreg., November.

Campbell, G.S., 1980, Long-wave radiation emission and absorption by domes of polyethylene shielded net radiometers: Pullman, Wash., Washington State University, 12 p.

Campbell, G.S., and Norman, J.M., 1998, An introduction to environmental biophysics (2d ed.): New York, Springer-Verlag, 286 p.

Carter, V., Reel, J.T., Rybicki, N.B., Ruhl, H.A., Gammon, P.T., and Lee, J.K., 1999, Vegetative resistance to flow in south Florida—Summary of vegetation sampling at sites NESRS3 and P33, Shark River Slough, November 1996: U.S. Geological Survey Open-File Report 99–218, 90 p.

Dingman, S.L., 2002, Physical hydrology: New Jersey, Prentice Hall, 646 p.

Dugas, W.A., Fritschen, L.J., Gay, L.W., Held, A.A., Matthias, A.D., Reicosky, D.C., Steduto, P., and Steiner, J.L., 1991, Bowen ratio, eddy correlation, and portable chamber measurements of sensible and latent heat flux over irrigated spring wheat: Agricultural and Forest Meteorology, v. 56, p. 1–20.

Dyer, A.J., 1961, Measurements of evaporation and heat transfer in the lower atmosphere by an automatic eddy-correlation technique: Quarterly Journal of the Royal Meteorological Society, v. 87, p. 401–412.

Engineering Toolbox, 2011, Water vapor-specific heat: Engineering Toolbox, accessed October 13, 2012, at http://www.engineeringtoolbox.com/water-vapor-d_979.html

Foken, T., 2008, The energy balance closure problem—An overview: Ecological Applications, v. 18, no. 6, p. 1351–1367.

Garratt, J.R., 1992, The atmospheric boundary layer: Cambridge, Cambridge University Press, 316 p.

Gash, J.H.C., and Dolman, A.J., 2003, Sonic anemometer (co)sine response and flux measurement I. The potential for (co)sine error to affect sonic anemometer-based flux measurements: Agricultural and Forest Meteorology, v. 119, p. 195–207.

German, E.R., 2000, Regional evaluation of evapotranspiration in the Everglades: U.S. Geological Survey Water-Resources Investigations Report 00–4217, 48 p.

Halldin, S., and Lindroth, A., 1992, Errors in net radiometry—Comparison and evaluation of six radiometer designs: Journal of Atmospheric and Oceanic Technology, v. 9, p. 762–783.

Hansen, K., 1959, The terms gyttja and dy: Hydrobiologia, v. 13, no. 4, p. 309–315.

Hostetler, S.W., 2009, Use of models and observations to assess trends in the 1950–2005 water balance and climate of Upper Klamath Lake, Oregon: Water Resources Research, v. 45, W12409, doi:10.1029/2008WR007295, 14 p.

Hubbard, L.L., 1970, Water budget of Upper Klamath Lake, southwestern Oregon: U.S. Geological Survey Hydrologic Investigations Atlas HA–351, 1 sheet.

Janssen, Kenneth D., 2005, Daily energy-budget and Penman evaporation from Upper Klamath Lake, Oregon: Portland, Oreg., Portland State University, M.S. thesis, 131 p.

Kann, P., and Walker, W.W., 1999, Nutrient and hydrologic loading to Upper Klamath Lake, Oregon, 1991–1998: Draft report prepared for Klamath Tribes Natural Resources Department and Bureau of Reclamation, Klamath Falls, Oreg., 48 p. plus appendices.

Kiehl, J.T., and Trenberth, K.E., 1997, Earth's annual global mean energy budget: Bulletin of the American Meteorological Society, v. 78, no. 2, p. 197–208.

Koberg, G.E., 1964, Methods to compute long-wave radiation from the atmosphere and reflected solar radiation from a water surface: U.S. Geological Survey Professional Paper 272–F, p. 107–136.

Lowe, P.R., 1976, An approximating polynomial for the computation of saturation vapor pressure: Journal of Applied Meteorology, v. 16, no. 1, p. 100–103. (Also available at http://dx.doi.org/10.1175/1520-0450(1977)016<0100:AAPFTC>2.0.CO;2.)

Massman, W.J., 2000, A simple method for estimating frequency response corrections for eddy covariance systems: Agricultural and Forest Meteorology, v. 104, p. 85–198.

Mauder, M., Liebethal, C., Gockede, M., Leps, J.P., Beyrich, F., and Foken, T., 2006, Processing and quality control of flux data during LITFASS–2003: Boundary-Layer Meteorology, v. 121, p. 67–88.

McCaughey, J.H., Lafleur, P.M., Joiner, D.W., Bartlett, P.A., Costello, A.M., Jelinski, D.E., and Ryan, M.G., 1997, Magnitudes and seasonal patterns of energy, water, and carbon exchanges at a boreal young jack pine forest in the BOREAS northern study area: Journal of Geophysical Research, v. 102, no. D24, p. 28977–29007.

Monteith, J.L., and Unsworth, M.H., 1990, Principles of environmental physics (2d ed.): London, Edward Arnold, 291 p.

Nash, J.E., and Sutcliffe, J.V., 1970, River flow forecasting through conceptual models—Part 1–A discussion of principles: Journal of Hydrology, v. 10, p. 282–290.

National Research Council, 2007, Hydrology, ecology, and fishes of the Klamath River Basin: Washington, D.C., The National Academies Press, 249 p.

Paw U, K.T., Baldocchi, D.D., Meyers, T.P., and Wilson, K.B., 2000, Correction of eddy-covariance measurements incorporating both advective effects and density fluxes: Boundary-Layer Meteorology, v. 97, p. 487–511.

Pearce, C.D., and Gold, L.W., 1959, Observations of ground temperature and heat flow at Ottawa, Canada: Journal of Geophysical Research, v.64, p. 1293–1298.

Sass, J.H., and Sammel, E.A., 1976, Heat Flow Data and Their Relation to Observed Geothermal Phenomena near Klamath Falls, Oregon: Journal of Geophysical Research, v. 81, p. 4863–4868.

Schedlbauer, J.L., Oberbauer, S.F., Starr, G., and Jimenez, K.L., 2011, Controls on sensible heat and latent energy fluxes from a short-hydroperiod Florida Everglades marsh: Journal of Hydrology, v. 411, p. 331–341.

Schotanus, P., Nieuwstadt, F.T.M., and deBruin, H.A.R., 1983, Temperature measurement with a sonic anemometer and its application to heat and moisture fluxes: Boundary-Layer Meteorology, v. 26, p. 81–93.

Schuepp, P.H., LeClerc, M.Y., Macpherson, J.I., and Desjardins, R.L., 1990, Footprint prediction of scalar fluxes from analytical solutions of the diffusion equation: Boundary Layer Meteorology, v. 50, p. 355–373.

Snyder, D.T., and Morace, J.L., 1997, Nitrogen and phosphorous loading from drained wetlands adjacent to Upper Klamath and Agency Lakes, Oregon, U.S. Geological Survey Water-Resources Investigations Report 97–4059, 67 p.

Stannard, D.I., 1997, A theoretically based determination of Bowen-ratio fetch requirements: Boundary-Layer Meteorology, v. 83, p. 375–406.

Stannard, D.I., Rosenberry, D.O., Winter, T.C., and Parkhurst, R.S., 2004, Estimates of fetch-induced errors in Bowen-ratio energy-budget measurements of evapotranspiration from a prairie wetland, Cottonwood Lake area, North Dakota, USA: Wetlands, v. 24, no. 3, p. 498–513.

Sturrock, A.M., Winter, T.C., and Rosenberry, D.O., 1992, Energy budget evaporation from Williams Lake—A closed lake in north-central Minnesota: Water Resources Research, v. 28, no. 6, p. 1605–1617.

Sumner, D.M., Wu, Q, and Pathak, C.S., 2011, Variability of albedo and utility of the MODIS albedo product in forested wetlands: Wetlands, v. 31, no. 2, p. 229–237.

Tanner, B.D., and Greene, J.P., 1989, Measurement of sensible heat and water-vapor fluxes using eddy-correlation methods—Final report prepared for U.S. Army Dugway Proving Grounds: U.S. Army, Dugway, Utah, 17 p.

Tanner, C.B., and Thurtell, G.W., 1969, Anemoclinometer measurements of Reynolds stress and heat transport in the atmospheric surface layer—Final report prepared for U.S. Army Electronics Command: Fort Huachuca, Ariz., Atmospheric Sciences Laboratory, 10 p.

Twine, T.E., Kustas, W.P., Norman, J.M., Cook, D.R., Houser, P.R., Meyers, T.P., Prueger, J.H., Starks, P.J., and Wesely, M.L., 2000, Correcting eddy-covariance flux underestimates over a grassland: Agricultural and Forest Meteorology, v. 103, p. 279–300.

Webb, E.K., Pearman, G.I., and Leuning, R., 1980, Correction of flux measurements for density effects due to heat and water vapour transfer: Quarterly Journal of the Royal Meteorological Society, v. 106, p. 85–100.

Western Regional Climate Center, 2012, Historical climate data: Western Regional Climate Center, accessed April 2, 2012, at http://www.wrcc.dri.edu/Climsum.html.

Wilson, Kell, Goldstein, Allen, Falge, Eva, Aubinet, Marc, Baldocchi, Dennis, Berbigier, Paul, Bernhofer, Christian, Ceulemans, Reinhart, Dolman, Han, Field, Chris, Grelle, Achim, Ibrom, Andreas, Law, B.E., Kowalski, Andy, Meyers, Tilden, Moncrieff, John, Monson, Russ, Oechel, Walter, Tenhunen, John, Valentini, Ricardo, and Verma, Shashi, 2002, Energy balance closure at FLUXNET sites: Agricultural and Forest Meteorology, v. 113, p. 223–243.

Winter, T.C., Rosenberry, D.O., and Sturrock, A.M., 2003, Evaporation determined by the energy-budget method for Mirror Lake, New Hampshire: Limnology and Oceanography, v. 48, no. 3, p. 995–1009.

Wood, T.M., Hoilman, G.R., and Lindenberg, M.K., 2006, Water-quality conditions in Upper Klamath Lake, Oregon, 2002–04: U.S. Geological Survey Scientific Investigations Report 2006–5209, 52 p.

Wood, T.M., and Gartner, J.W., 2010, Use of acoustic backscatter and vertical velocity to estimate concentration and dynamics of suspended solids in Upper Klamath Lake, south-central Oregon— Implications for *Aphanizomenon flos-aquae*: U.S. Geological Survey Scientific Investigations Report 2010–5203, 20 p.